THE SCARIEST PLACE ON EARTH

BOOKS BY DAVID E. FISHER

THE SCARIEST PLACE ON EARTH

Eye to Eye
with Hurricanes

David E. Fisher

RANDOM HOUSE

NEW YORK

*Grateful acknowledgment is made
to the following for
permission to reprint previously published material:*

FRED AHLERT MUSIC GROUP AND CPP/BELWIN, INC.: Three
lines from "Stormy Weather" by Harold Arlen and Ted
Koehler. Copyright © 1933 (renewed 1961) by EMI Mills
Music, Ted Koehler Music, and SA Music. All rights for
the extended term of copyright administered by Fred Ahlert Music Corporation on behalf of Ted Koehler Music
Company. Harold Arlen share administered by SA Music.
Rights throughout the world excluding the United States
are controlled by CPP/Belwin, Inc. International copyright secured. Made in the USA. All rights reserved. Reprinted by permission.

FAMOUS MUSIC CORPORATION: Three lines from "That Old
Black Magic" by Johnny Mercer and Harold Arlen. Copyright © 1942 by Famous Music Corporation. Copyright
renewed 1969 by Famous Music Corporation. Reprinted
by permission.

WARNER/CHAPPELL MUSIC, INC.: Three lines from "Dance
Little Lady" by Noel Coward. Copyright © 1928 (renewed) by Chappell Music Ltd. All rights for the United
States controlled and administered by Chappell & Co. All
rights reserved. Reprinted by permission.

WILLIAMSON MUSIC: Three lines from "A Puzzlement" by
Richard Rodgers and Oscar Hammerstein II. Copyright ©
1951 by Richard Rodgers and Oscar Hammerstein II.
Copyright renewed. Williamson Music owner of publi-
cation and allied rights throughout the world. International copyright secured. All rights reserved. Reprinted
by permission.

Library of Congress Cataloging-in-Publication Data
Fisher, David E.
The scariest place on Earth: eye to eye with hurricanes
David E. Fisher.
p.   cm.
Includes bibliographical references.
ISBN 0-679-42775-9
1. Hurricanes.   2. Hurricane Andrew, 1992.   I. Title.
QC994.F57   1994   363.3′492 — dc20   94-1810

Manufactured in the United States of America
Book design by Tanya M. Pérez
2   4   6   8   9   7   5   3
First Edition

This book is for Leila L.
Who else?

How utterly we have forsaken the Earth, in the sense of excluding it from our thoughts. There are but few who consider its physical hugeness, its rough enormity. It is still a disparate monstrosity, full of solitudes and barrens and wilds. It still dwarfs and terrifies and crushes. The rivers still roar, the mountains still crash, the winds still shatter. Man ... somehow ... has managed to shut out the face of the giant from his windows. But the giant is there, nevertheless. ...

WALLACE STEVENS, *Letters*

Andrew was the worst disaster ever to hit the United States. It was as if a nuclear bomb exploded over South Florida.

FEDERAL EMERGENCY MANAGEMENT AGENCY, August 24, 1993

PROLOGUE

From *USA Today,* Monday, August 24, 1992:

The scariest place on earth is directly in the path of an onrushing
category 5 hurricane. That may sound simplistic or even blasphe-
mous to people who have read about the horrors going on in Bosnia
or Kurdish Iraq, or who have memories of enduring the Blitz in
London or the fire-bombing of Dresden, or who managed to survive Hiroshima or Nagasaki. People who were trapped by the California earthquakes of 1989 or 1994 or who ran from the Mount
Pinatubo eruption in 1992 may think they know all there is to know
about being frightened. With sympathy and respect to them all, let
me tell you about hurricanes.

"Above the howling wind and the driving rain, the villagers of
Manpura Island could hear an unholy roar welling up from the Bay

ix

of Bengal," *Time* magazine began its report of November 30, 1970, telling about the hurricane that killed half a million people—more than were killed at Hiroshima and Nagasaki and Dresden and London *combined.* One year earlier, four out of every five houses in Pass Christian, Mississippi, were destroyed by Hurricane Camille. Twenty years later the Caribbean islands of Vieques and Culebra were annihilated by Hurricane Hugo. To the people standing virtually naked and alone in the paths of those storms, there could have been no scarier place on earth.

Nor is the danger from hurricanes confined to tropical places, or even to coastlines. In 1900 a hurricane devastated Galveston, Texas, killing more than five thousand people, and kept right on going. It maintained its intensity right up through the breadbasket of the United States, took a right turn over the Great Lakes, swamped the northeastern states and headed out over the Atlantic, smashed into Europe, and kept right on going, finally disappearing into the wastes of Siberia.

The fury of the hurricane is enormous, almost beyond belief. The energy released in just one decent-sized hurricane is as much as that in 500,000 atom bombs. When we demolished Bikini Island with a thermonuclear-bomb test, the explosion lifted 10 million tons of water into the air; a hurricane over Puerto Rico drenched the island with 2.5 *billion* tons of water. In 1954 Hurricane Hazel tore a two-story, reinforced-concrete building loose from its foundations and flung it a few hundred yards through the air. Along with the ruins of the structure was found an ebony cup, engraved "Made in Haiti"—the storm had carried it a thousand miles before dropping it in North Carolina.

The Great Hurricane of 1780 demolished the West Indies. The commander of the British fleet that was stationed there ordered his ships to sail out of harm's way as it approached; after it left he sailed back, and later wrote to his wife:

> It will be difficult for you to realize my surprise and grief when
> I saw the devastated condition of the island and the destructive

force of the hurricane. The strongest buildings and a number of houses, the majority of which were made of stone, yielded to the fury of the wind. The entire fort was destroyed and many heavy cannons were carried a distance of more than 100 feet. If I had not seen all these things myself it would have been difficult for me to believe it. More than 6,000 people died and whole settlements were destroyed.

In recent years we haven't heard much about the terror of hurricanes, probably because atmospheric conditions in the past couple of decades have diminished their frequency. But, as we shall see, the immediate future looks more like a return to the old days, and, with global warming, a return with a vengeance. The old days were terrible enough, as told in a nineteenth-century report from a Russian traveler:

> The deafening noise of the wind changed to bellowing or, to be exact, to a distant roar which is difficult to express in words, and now the lightning illuminated the entire space between clouds and earth and this continued for half a minute. It appeared as if a huge fire erupted in flames next to the houses, which went up immediately.
>
> Immediately after this terrible phenomenon, shown by the lightning, the hurricane again started with such a terrific force that it carried everything that could not withstand its pressure. It is not only that solid buildings started shaking but even the surface of the earth was vibrating. It was not possible to hear the peal of thunder at this time. The terrible howl and roar of the wind, the noise of the sea waves threatening total destruction of the city, the noise due to the fall of walls, windows, fencing, etc. cannot be conceived. . . . It is difficult for those who live far from the terrible, agonizing situation to visualize the condition and suffering of the people living in this city.

In this book I want to try to tell you about hurricanes — what they are, how they form, what damage they can do, and how we might modify or avoid them — and about the scariest place in the world: Miami, Florida, on August 24, 1992. . . .

AUTHOR'S NOTE

A note about units: the metrication of America is proceeding slowly, and I have temporized. Temperatures are given in degrees Centigrade, but I have never yet seen a weather report or heard a conversation in which speeds are given in units of meters per second, and so I have used miles per hour in this book. Wave heights are also given in feet rather than in meters, with apologies to those who know better. All other units are defined where used.

Names of historical characters and places are spelled as in the original documents cited.

ACKNOWLEDGMENTS

I am grateful for informative discussions with Drs. Gote Ostlund and Naomi Sturgi of the Rosenstiel School of Marine and Atmospheric Science of the University of Miami, and to Dr. Helen Albertson and Kay Hale, librarians therein. Any inaccuracies in this text are due solely to my own misinterpretations.

CONTENTS

THE SCARIEST PLACE ON EARTH

1

FIRST ENCOUNTERS

Don't know why
There's no sun up in the sky,
Stormy weather...

—from "Stormy Weather,"
by Harold Arlen and Ted Koehler

1

During hurricane season *The Miami Herald* routinely reports the path of all tropical depressions that turn into tropical storms, watching to see if they develop into full-scale hurricanes. Nobody pays any attention. On Wednesday, August 19, 1992, midway through the twenty-sixth consecutive hurricane season in which no storm had touched Miami, an unobtrusive report of the year's first potential hurricane appeared in a small box on a back page: "The first tropical storm of 1992 maintained its strength Tuesday evening with winds of about 50 mph, but was moving more slowly. A reconnaissance flight into Andrew is scheduled today...."

Of all hurricane reports I had ever seen, this was among the least likely to worry us. The normal path of a hurricane is to the west across the Atlantic, and then around to the north as it approaches the continental United States. For a hurricane to be a threat to Miami it has to start out far to the south, below the islands of the West Indies. Andrew had started at about latitude 15 degrees north and had already moved up to nearly 17 degrees north; its direction would take it far to the east and north of Miami. If it hit the coast at all it would probably be somewhere around the Carolinas.

I turned the page to read about events of more local concern.

3

2

One winter day in 1965 my wife, Leila, and I were sitting in our living room in Ithaca, New York, with the wind blowing and the kids sniffling with colds and the snow battering our windows, and we looked at the map. We had just decided to leave Cornell to go to the marine lab at the University of Miami.

"There," my wife said.

I looked at where she was pointing.

"That's where I want to buy a house," she said.

She was pointing to Key Biscayne, a chunk of land jutting out from the Miami shoreline, sitting by itself a mile offshore. You could almost see the white sand on the map, almost feel the warm sea breezes.

"Uh-uh." I shook my head. "There's gonna be a hurricane every few years down there. You can't live out in the middle of the water."

"Don't be silly," Leila explained, so when we went down to Florida and began searching for a new home, we looked at Key Biscayne first, and I was happy to find that Richard Nixon had built a vacation retreat there and that the real estate prices had tripled and we couldn't afford it. Instead we bought a house in a suburb to the west of Miami, a place called Kendall. A month later I had to leave town on business. When I called a few days later my wife was worried because a hurricane was approaching and was due to hit the next day. I told her she had nothing to worry about. The only real danger from a hurricane, I explained ignorantly, is the storm surge: the combination of low barometric pressure and strong winds lifts the water right out of the ocean basin and pours it onto the shore, just as if someone were carelessly jerking around a bucket of water. It sloshes over the sides and can rise into ten-foot or even higher waves, which might flood your house or even knock it down. But this happens only to the people who live on the beaches, and we were a good ten miles inland and out of danger. "See," I said, "that's why I didn't want to buy on Key Biscayne. You've got nothing to worry about."

She didn't believe me, of course, and our new neighbors came over and showed her how to put down the metal shutters. The house was locked up tight when the hurricane hit, and, as I had said, it was nothing worse than a bad storm. When I returned, Leila nearly admitted I had been right.

3

Hurricanes, by definition, occur only in the Atlantic. In the Pacific they are known as typhoons. (In 1687 a sailor named William Dampier, who had experienced typhoons in the China Sea and hurricanes in the Indies, recognized that both these storms are the same.) The first such storm to imprint itself on our history, in fact, occurred in the Pacific.

In 1274 Genghis Khan's grandson Kublai Khan stood at the head of the greatest empire the world had ever known. Genghis and his son, Ögödei, had marched from the Asian steppes into Europe and the Mideast, conquering the assembled armies of Christendom and Islam. Turning his attention to the east, Kublai Khan had conquered China and Korea, and on November 19th, 1274, he invaded Japan with a warfleet of hundreds of ships carrying perhaps forty thousand warriors. They landed on the small offshore islands of Tsushima and Iki, and then advanced toward the main island, invading Japan proper at Hakata Bay.

The Japanese were in the midst of their shogun era, and had settled down from a warlike, internally feuding people to a comfortable and prosperous one. They were in no condition to repel the Mongol hordes, who dispatched the numerically inferior army sent to oppose them. Then, as the Khan's generals were preparing to march inland, a storm began whipping up the seas. The army lay exposed on the beach, and the Korean admirals who handled the fleet convinced the generals to reboard the ships and ride the storm out at sea. They had great confidence in their ships, which were more advanced than any others in the world at that time, featuring as many as thirteen separate watertight compartments that could be sealed off, and a stern rudder (just then coming into use in Europe).

Unfortunately for them this was no normal storm, but a typhoon. The Korean navy with its finest ships was no match for the fury whipping up the waters. Hundreds of ships were sunk, thousands of men died, and the remainder lost all thought of fighting. The Mongol invading force was destroyed, and the survivors sailed home as best they could in their crippled ships.

The mighty Khan was not accustomed to such rebuffs from man or god, and seven years later he struck again, this time with the largest invasion force in history—reportedly a thousand ships carrying nearly two hundred thousand soldiers. But he hadn't learned from the previous disaster: It hadn't been the opposing Japanese but the storm that had defeated him, and now in their efforts to heed the Khan's rampaging shouts for revenge his shipbuilders hurried their construction and took shortcuts, ensuring that the fleet couldn't possibly survive another storm.

But perhaps Kublai thought such bad luck couldn't strike twice, a hope that survives to the present day among hurricane victims and is as vain now as it was then. He divided his forces into two parts, and landed both forces on the Japanese coast. The Japanese, fighting furiously in defense of their homeland, kept the Mongols confined to the beachhead all through the summer, fighting a steadily losing series of battles from June through July and on into August.

The fighting continued until August 15th, when once again the winds roared out of the east, smashing the invading fleet against the shore, drowning half their men, leaving the remainder dazed and battered onshore where the Japanese, who were familiar with typhoons, then came flowing out of their prepared bunkers and slaughtered them. From that time onward the "divine wind," or "kamikaze," has been revered in Japanese lore. The Japanese air force tried to emulate it in 1945, but a fleet of suicidal airmen proved to be no substitute for a full-scale typhoon.

. . .

On the other side of the world, the European/Mediterranean civilization knew nothing of the existence of such terrible storms, since they begin innocently enough as undeveloped weather systems off the western coast of Africa and travel from there west across the Atlantic. If they don't hit the continent of America and dissipate, they curve north and east and die out in the northern Atlantic. Rarely have they survived long enough to hit Europe, and when they have they've been regarded as severe but normal storms. Not until the age of oceanic exploration, when the first men to travel across the Atlantic ran into these strange phenomena, were hurricanes perceived as anything different.

It is an old wives' tale that Columbus sailed west in 1492 with his sailors fearing they would fall off the edge of the world. Everyone who cared at all about such things already knew that the world was round. "I have always read," Columbus wrote, "that the world, both land and water, is round; the arguments and experiments recorded by Ptolemy and every other writer on the subject supply proofs of the matter, in the form of lunar eclipses and other observations made from east to west in addition to the elevation of the pole from north to south."

It was a Greek philosopher, Eratosthenes, who had proved the roundness of the earth nearly two thousand years earlier. He had done more: he had calculated just how big the earth's sphere was, by adding an ingenious application of trigonometry to a unique observation.

He had noticed that every year on June 21, a straight tree in the southern Egyptian city of Syene (where he was born) cast no shadow at noon. In Alexandria, 480 miles to the north, he saw that a straight tree there did cast a shadow at noon on that date. He realized that the lack of shadow meant the sun was directly overhead at that time, and if it was not also overhead at Alexandria, it could only be because of the curvature of the earth (see Figure 1-1).

Drawing straight lines from both cities to the hypothetical center of the earth defines the angle A', and since alternate interior angles of two parallel lines are equal, it also defines the angle A. Angle A

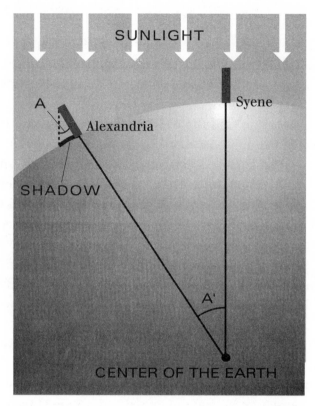

SUNLIGHT

A

Alexandria

Syene

SHADOW

A'

CENTER OF THE EARTH

FIGURE 1-1

can be calculated by measuring the length of the shadow and using a little basic trigonometry. The radius of the earth — the distance from either city to the center — can then be calculated from the same trigonometric laws, knowing the angle A' and the distance from Syene to Alexandria. Eratosthene's answer came out to be just less than four thousand miles, and (knowing that $C=2r$) the earth's circumference was estimated at 24,000 miles (although there is some uncertainty because he used the Greek unit, stadium, instead of the mile, and we don't know the exact dimension of the stadium). The correct values are about 25,000 miles for the circumference and 3,950 miles for the radius.

The general result of Eratosthenes's calculation — that the world was spherical — was remembered by scholars and sailors when the world fell into the Dark Ages, but the details were lost. When

Europe surfaced into the Renaissance hundreds of years later, its scholars were more likely to turn to mythological and biblical sources for enlightenment instead of to mathematics. For example, about fifteen hundred years after Eratosthenes, nearly a hundred years before Columbus sailed, Cardinal Pierre d' Ailly was chancellor of the University of Paris. He knew enough of Greek science to know the world was round, but not enough to know its size. Ignoring mathematics, he found in the biblical Apocrypha the statement that one-seventh of the earth's surface is covered with water. From what he knew of the dimensions of the land, he calculated that the ocean which separated Spain from the Indies couldn't be more than four thousand miles wide. In truth, it is roughly ten thousand. He was wrong about the fraction of the earth covered by water, the total area of the earth, and the Bible as a source of scientific knowledge. But his mistakes were an inspiration to Columbus, for in the margin of Columbus's copy of d' Ailly's *Imago Mundi,* still preserved at the Cathedral of Seville, can be found his handwritten note: "The end of Spain and the beginning of India are not far distant but close, and it is evident that this sea is navigable in a few days with a fair wind."

Once Columbus became interested in the possibility of sailing to the East by steering west, he investigated all the scholarly works that treated the problem. Ferdinand Columbus's biography recalls his father explaining how Aristotle had shown that the passage from India to Spain shouldn't take more than a few days. Seneca, writing at about the time of Christ, repeated this as a simple and well known fact, and in the world's first book of geography, Strabo pointed out that "the Ocean encompasses the whole earth, it washes India on the East and Spain on the West, and but for its vast extent, one might sail directly from the one of these countries to the other." Ferdinand Columbus goes on to say that both Pliny and Solinus wrote of the one ocean that surrounds the earth, and that a careful reading of Marco Polo implies the nearness of Spain and India. Moreover, "after a course of westerly winds," fragments of wood, whole logs, and indeed even a couple of human corpses "who looked quite unlike Christians" had been known to wash up on the western shore of the Azores, indicating the existence of inhabited land to the west, beyond the horizon.

Indeed, Columbus had noticed something similar on a voyage he made to Ireland in 1477. As related by naval historian Samuel Eliot Morison, he saw two boats drifting off the coast. In the boats were a dead man and woman "of extraordinary appearance." The term evidently referred to appearances other than mid-European Caucasian, and from this Columbus deduced that they must be Chinese who had drifted to Ireland across the uninterrupted sea. (Morison surmises that they might have been Lapps or Finns, drifting down from the north.)

Aside from his tendency to fasten on any evidence, however slight, that showed the Atlantic to be no more than a large pond, Columbus was in good company in the firm knowledge that the earth was spherical. Sailors in particular knew this: they navigated the seas by using compass readings and solar sightings, making use of the apparent position of the sun as seen by travelers on a spherical earth, and were well acquainted with the sights of ships gradually disappearing over the horizon. Their charts, dating back to the time of Ptolemy, were known to be flat representations of a spherical earth.

What they had to fear was more terrifying than the chimera of reaching and falling off the edge of the earth: what they had to fear was the unknown. For who knew what lay beyond the horizon, beyond the sight of land? Who knew what evil lurked in the depths of the sea?

They were quite right to be afraid. Although they never fell off the edge of the earth and they were not devoured by mythical sea dragons, nor seduced onto rocks by lasciviously evil mermaids, nor trundled up by angry demons, the seas they sailed into were the provinces of storms more terrible than any European sailor had ever imagined.

. . .

If Columbus had encountered a hurricane at sea during that voyage of 1492, if he and his three little ships had disappeared without a trace—as indeed they would have, in such a case—how different the history of the world would be! Fortunately, there was no hurricane that year, nor the next, nor the year after that. It wasn't until 1495 that Europeans first saw a hurricane; it was not one that they would soon forget:

It was one of those awful whirlwinds which occasionally rage within the tropics, and were called by the Indians "furicanes," or "uricans," a name they still retain with trifling variation," the early American writer Washington Irving reports in his biography. "After mid-day a furious wind sprang up from the east, driving before it dense volumes of cloud and vapor. Encountering another tempest of wind from the west, it appeared as if a violent conflict ensued. The clouds were rent by incessant flashes, or rather streams of lightning. At one time they were piled up high in the sky, at another they swept to the earth, filling the air with a baleful darkness more dismal than the obscurity of midnight. Wherever the whirlwind passed, whole tracts of forest were shivered and stripped of their leaves and branches: those of gigantic size, which resisted the blast, were torn up by the roots and hurled to a great distance. Groves were rent from the mountain precipices, with fast masses of earth and rock, tumbling into the valleys with terrific noise, and choking the course of rivers. The fearful sounds in the air and on the earth, the pealing thunder, the vivid lightning, the howling of the wind, the crash of falling trees and rocks, filled every one with a fright; and many thought that the end of the world was at hand. Some fled to caverns for safety, for their frail houses were blown down, and the air was filled with the trunks and branches of trees, and even with fragments of rocks, carried along by the fury of the tempest. When the hurricane reached the harbor, it whirled the ships round as they lay at anchor, snapped their cables, and sank three of them with all who were on board. Others were driven about, dashed against each other, and tossed mere wrecks upon the shore by the swelling surges of the sea, which in some places rolled for three or four miles upon the land. The tempest lasted for three hours.

When it had passed away and the sun again appeared, the Indians regarded each other in mute astonishment and dismay. Never in their memory, nor in the traditions of their ancestors, had their island been visited by such a storm. They believed that the Deity had sent this fearful ruin to punish the cruelties and crimes of the white men; and declared that this people had moved the very air, the water, and the earth, to disturb their tranquil life and to desolate their island.

Well, Mr. Irving did have a tendency to get a bit carried away by the vehemence of his own prose, sometimes to the point of obvious exaggeration, such as the statement that "Never in their memory, nor in the traditions of their ancestors, had their island been visited by such a storm." This is hardly likely unless the entire tribe suffered from Alzheimer's, for not only had the Indians seen hurricanes before, they had invented the word.

When these peoples had first reached the Caribbean, Central America, and the Gulf region some ten thousand years earlier, they had found the hurricanes waiting. They came across the Bering Strait at the end of the last ice age, when the ice had receded enough to allow passage and the seas had not yet risen to cut Asia off from America. They came inching down through the continent, slaughtering everything in their wake, bringing about the largest mass extinction of species since a giant meteoroid hit the earth some 65 million years earlier. They came pressing down through the snowstorms and the barren landscapes until they found the verdant hills and valleys of the tropics, and there—to the names of the myriad gods and goddesses they had brought with them—they added the name of the mightiest god of them all.

They separated into disparate tribes and clans, each settling into its own niche, each with its own word for the great god of storm and fury, and for the terrible evidence of his prowess. The Galibi called him Yuracan, the evil spirit; the Maya named him Jurakan, and sacrificed living people to him, begging him not to bring the terrible winds. In the Honduras he was Kukulkan, the supreme god of gods. To avert catastrophe, or at least to ameliorate the storm's destruction, they would blow trumpets into the air when the winds rose, or cast straws against the tempest; they would beat on drums and huddle in stone houses if they had them, or flee to the forests if they did not. The Spaniards, of course, were more sophisticated. Since the hurricanes peaked in intensity in October, the month of Saint Francis of Assisi, they imitated the knotted cord around his waist by hanging a little rope with three knots in the doorways of their houses during the hurricane season. And the Indians and the Spaniards were just about as effective in averting the hurricane's destruction as we were in Miami in 1992.

4

Columbus set sail for Spain in the *Niña*, accompanied by Juan Aguado in the *Santa Cruz*, the only ships that were spared by the hurricane of 1495. Because of the loss of his other ships these two were impossibly crowded: the *Niña* had carried less than twenty-five men on her original voyage across the Atlantic, and the *Santa Cruz* was the same size; together they now bore 225 whites and thirty Indians (and an untold number of syphilis spirochetes, which prospered in Europe better than Columbus ever did.)

The difficulty facing Columbus in the Indies was that he had found little sign of gold and less sign of China. There was some gold, to be sure, just enough to excite the Spaniards' passion. The natives they first met were wearing gold trinkets, ear plugs, necklaces, and bracelets, which titillated the Spanish imagination and drew them into the interiors of the islands to search for the source of all that treasure, for rich hidden mines.

But these gold mines did not exist. The gold in those islands was present in tracer deposits only. Enough washed down through the streams for the Indians to use for jewelry, but what was enough gold for a few thousand Indian trinkets was not enough to turn the Spanish galleons into heavily laden armories of wealth.

Columbus tried to counter the complaints brought back by his men to the Spanish king and queen with gifts of as much gold as he could squeeze out of the islands, but this was not enough to satisfy Their Majesties. So he also brought back thirty natives to convince his sponsors of another source of riches: slavery. These natives, he told King Ferdinand and Queen Isabella, were ignorant of the one true Lord. It was one of the things he loved about them, for being without religion they were as innocent as babes and could easily be converted to the true faith. (He never considered the native religion, which encompassed a wide spectrum of gods and goddesses, as worthy of consideration.) The other thing he loved about them was that they were friendly and guileless, unassuming and unwarlike; in short, perfect for slavery. He evidently saw no conflict between the desire to save them for Christianity and his wish to use them as slaves. Their Majesties could, he said, at the same time save their souls for God and use their bodies for Spain.

One should note that he was not behaving in an inhuman manner in thinking of the Indians as a source of slaves. On the contrary, he was all too human. Slavery was not invented by the white Europeans, nor is it a vice peculiar to them. Everyone practiced it, wherever and whenever they could: the ancient Greeks and Romans, as well as all other ancient peoples. (The Hebrews included a provision that a man held as slave must be set free after seven years; in all other cultures, slavery was for life.) In more modern times slavery was still ubiquitous. African blacks took their captives as slaves, and even initiated wars specifically to gather slaves, either for their own use or to sell to the white men. Arabs traded in slaves, and many of the Indian tribes of the New World kept Spanish castaways, victims of hurricanes, as slaves. In the days before we had machinery to work for us, the temptation to take a weaker ("inferior") person as a slave must have been irresistible.

The Spanish sovereigns in particular would not be the ones to cavil at Columbus's suggestion. They had just finished rounding up the Jews, dispossessing them of everything they owned, and expelling them from their kingdom. This, after years of trying to save their souls for God by burning their bodies at the stake, is evidence enough that rulers such as these were not likely to make too fine a moral point out of enslaving dark-skinned people.

Nor indeed did they, accustomed as they were to the joys of the Inquisition, when not only Jews but Protestants, Moors, and Gypsies, intellectuals and atheists, were all burned at the stake to be rid of heresy. A story was told in Spain during the 1490s about a certain nobleman who died and went to heaven. The Lord accompanied him down the long halls of the heavenly castle, pointing out to him the various rooms from among which he could choose his eternal paradise. "In here," the Lord said, "are the Chinamen and Indians who call themselves Buddhists. Along here," He said, continuing on down the hall, "are the Lutherans, down here are the Protestants, here the Moors, and there the Russian Orthodox. . . ." And then as they came upon the next room the Lord rose up on His tiptoes and put His finger on His lips. "Shh," He whispered, and they passed by that door quietly and went on. "Here we have the Jews," the Lord said as they came upon the next room, at which point the nobleman begged to interrupt Him and asked, "What about that room back there? The one we tiptoed past?"

"Shh," the Lord whispered again. "That room is for the Catholics. They think they're the only ones up here."

Well, I suppose they didn't really tell that story in Spain in the 1490s, as Christian arrogance was no laughing matter. Certainly not for the Jews who were burned, nor for the West Indians who would be enslaved. Of course according to the prevailing point of view, it was all done in their best interests. As Irving says, "Nor were the interests of the unhappy natives forgotten by the compassionate heart of Isabella . . . [who] always consented with the greatest reluctance to the slavery even of those who were taken in open warfare; while her utmost solicitude was exerted to protect the unoffending part of this helpless and devoted race. She ordered that the greatest care should be taken of their religious instruction . . ."

Isabella's "compassion" didn't seem to help; the Indians were more than enslaved, they were annihilated. For Columbus was wrong to assume their docility and eagerness to please would make them suitable candidates for servitude. When they were enslaved they either fought back or withered away; either way, they died. Of the 300,000 Tainos indians inhabiting Hispaniola when Columbus first made landfall there in 1492, one third were killed by 1496. In 1508 a census showed only sixty thousand still alive, in 1512 the number was twenty thousand, and by the middle of that century the Spanish governor of the island reported that fewer than five hundred Indians were to be found. Today none remain.

But at the time Columbus's arguments pleased the Spanish sovereigns, and they sent him back twice more to the Indies to find a path to China, to find the Indians' hidden gold mines, and to bring back armies of slaves. In order to keep the peace among the colonizers, however, Ferdinand ordered him not to interfere with the new administration of the islands, which was to be set up without him. On his fourth and final voyage, in 1502, Columbus's main mission was to get to China, but he was forced to take a detour. His flagship on that trip, *La Capitana*, proved untrustworthy on the transatlantic voyage, and so he put into the port of Santo Domingo to attempt to replace her. He also wanted to find shelter there from a hurricane that he felt coming.

I use the phrase "felt coming" because I can't think of a better one. The signs he listed as indicative of the impending storm are less than sure: cirrus clouds scudding along ahead of an "oily" rolling swell of ocean waters otherwise suspiciously smooth, like glass; light winds with sudden sharp gusts; large sea creatures (seals, manatees, and unidentified shapes) coming to the surface, apparently deserting the turmoil of the deep waters; and perhaps most telling, sharp pains in his rheumatic or arthritic bones. Somewhat less than a scientific basis, especially for a sailor who had seen only one hurricane (possibly two) before, but as it turned out he was right.

Knowing he was forbidden to land in Santo Domingo, he hove to offshore and sent one of his officers to ask for sanctuary. The governor, Nicolás de Ovando, refused his request and told him to shove off, in more or less those words.

When he received the curt reply, Columbus sent his officer back with a letter assuring Ovando that a hurricane was truly coming, and implored him at least to make sure his own ships were safe.

Ovando laughed at the warning, while reading the letter aloud to his grinning court. His fleet was ready to sail to Spain with a fortune in gold—including one ship with the largest single mass ever seen on these islands, and another laden with Columbus's personal fortune of four thousand pieces—and he was not going to let Columbus interfere. Especially not with such nonsense. Looking out on the harbor, they could all see a day fair and tranquil, with not a hint of a storm in the sky.

Ovando's fleet sailed on the morrow, as Columbus took his own group of ships along the coast and found an inlet to shield them from the easterly winds he was expecting. Two days later, the storm struck:

> The baleful appearance of the heavens, the wild look of the ocean, the rising murmur of the winds, all gave notice of its approach. The fleet [Ovando's] had scarcely reached the eastern point of Hispaniola when the tempest burst over it with awful fury, involving everything in wreck and ruin. The ship aboard which . . . were a number of the most inveterate enemies of Columbus was swallowed up with all its crew, and with the celebrated mass of gold, and the principal part of the ill-gotten treasure gained by the miseries of the Indians. Many of the ships were entirely lost . . . and only one was enabled to continue her voyage to Spain. That one was the weakest of the fleet, and had on board the four thousand pieces of gold, the property of the Admiral.

Nineteen ships disappeared with all hands; another half-dozen were lost, but from these somehow a few sailors survived; and four managed to get back into harbor before they sank. Altogether, more than a million dollars in gold sank under the waves. And the *Aguja*, with Columbus's gold, sailed on to Spain.

The man was a demon, everyone agreed, who had obviously made a pact with the devil. Columbus himself saw it a bit differently: "He was deeply impressed with awe, and considered his own preservation as little less than miraculous. Both his son Ferdinand

and the venerable historian Las Casas, looked upon the event as one of those awful judgments which seem at times to deal forth temporal retribution."

5

These were the first hurricanes encountered by white men, but far from the last. As well as can be determined, hurricanes were about as frequent in the early centuries of Atlantic exploration as they are today. Yet it is hard to be certain, because most of the ships that encountered hurricanes were lost, without survivors, and what destroyed them can only be conjectured. Slowly, some of the most rudimentary facts about the storms found their way into nautical lore. That the winds always blew counterclockwise, for example, and that although the winds themselves blew furiously, the actual passage of the storm was at a rate of only ten to twenty knots. Putting these two facts together meant that a wise captain could run out of the path of a storm if he could learn to recognize its portents quickly enough. If the first winds came out of the west, he knew the storm must be northward; if they came from the east, it would be blowing up from the south. Since a ship could make speed at nearly the rate at which the hurricane was approaching or receding, the captain could steer away from it or tack across its path and get far from the center's turmoil. At the very least, he could avoid catching up and running into it.

But how to recognize the signs? A twinge in one's arthritic joints? Seals on the surface? An oily swell? Don't forget, if you alter course and delay the voyage by a few days, and another twenty or thirty slaves die because of the delay, your ship's owners tend to get nasty. Turn aside too soon and you'll lose your command; turn aside too late and you'll lose your life as well. Those were not easy days to sail the seas.

One of the more horrifying stories tells of Valdivia, an officer under Balboa, who, bound for Hispaniola, was caught by a hurricane near Jamaica in 1519. His caravel lay northeast of the approaching storm, and instead of trying to cut east or west to get out of its path, the captain—being wholly ignorant of the movements of hurricanes—set his sails head-on and tried to run from it. Useless. The storm blew the caravel along like a child blowing on a toy sailboat while swimming behind it; the storm blew it out of the Caribbean, into the Gulf of Mexico, shattered it on a protruding reef, and roared on.

The crew, battered and dazed, had no idea where they were. The ship was broken and leaking badly, but their only choice was to sail in it or die where they were. With no idea of which way to go, they simply picked a direction at random—west—and stuck to it. In the next two weeks six of the eighteen men who survived the hurricane died. Luckily the survivors hit a strange new coast. Unluckily, it was what is now called Yucatan, where the natives who found them staggering onto the beach were not the docile, welcoming Indians Columbus had encountered but a strong and proud tribe. Valdivia made himself known as the leader of the party, and as one who expected certain privileges.

He got them. The natives paraded their find through a city that amazed the Spaniards, hidden as it was in the empty vastness of the New World. The streets were lined with stone, and the people were far from naked: they were dressed in clean white cloth and, but for their darker color, could have passed for people on any street in Spain. The city was comprised of gigantic stone buildings, taller than any the Spaniards had seen, topped with pyramidal structures pointing to the sky. Obviously these people shared a metaphysical orientation with Europeans, who built spired cathedrals pointing toward heaven.

Yet they were not exactly the same. They locked their captives in cages, except for Valdivia, whom they evidently recognized as someone special. He was allowed his freedom, and was fed handsomely and treated courteously. Though they were insistent upon having their own ways: they painted him a bright blue.

Four other captives were also painted blue, the remainder being splashed with black and white. The blues, like Valdivia, were allowed their local freedom; the black-and-whites were kept in the cages. But all were otherwise treated well, fed and taken care of. One of the black-and-whites, a former priest named Aguilar, learned to speak the natives' language, after a fashion. And then one morning a procession of obviously important natives came for Valdivia, bringing fine clothing. They dressed him ceremoniously and gave him a potion to drink, which he found delicious, and which made him smile broadly and walk a bit unsteadily. They dressed him in garments that seemed to be made of spun gold, and crowned him with a birdlike headdress. And then they led him away.

They took him to a structure that at first seemed like a solid wall, but as they approached Valdivia saw that it was pyramidal. Steps led up toward the top which disappeared, it was so high. They gestured for him to take the first step, and then the second, and then another and another. They climbed into the sky, a vast ceremonial procession bedecked in brightly colored bird feathers and swirling gold. As they passed higher and higher, great masses of people came behind. They filled the steps on which they climbed, flowing upward like a silent, misguided river.

At the top they were so high that patches of clouds swirled around, and between them Valdivia could see out over the jungle below, over the city of stone. Gently they guided him forward to a stone altar, and then out of the clouds that spun around them appeared an old man, a nightmarish figure streaked with blood. Valdivia started back, but strong hands caught him, turned him around silently, and bent him backward over the altar. The last thing he saw was the bright blue sky of Mexico, the swirling white clouds, and then a sudden flash of darkness as the blood-streaked priest held a black flint knife high over him. And then the knife plunged down.

It plunged into his chest, tearing away his flesh. The priest reached in and ripped out his beating heart, held it high over his head as the crowds below shouted and cheered, sang and prayed, and offered the living heart to their living gods.

Thus did white men first find the Mayan civilization.

Thus also did they find Bermuda, which had first been charted by Juan de Bermúdez in 1511. It was known then not by his name but as the *Ya de Demonios,* the Islands of Devils, because of the stormy weather and the reefs that surrounded the chain of islands. A contemporary account titled *Discovery of the Bermudas otherwise called the Ile of Divels* states: "For the Ilands of the Bermuda, as every man knoweth that hath heard or read of them, were neuer inhabited by any christian or Heathen people, but euer seemeth and reputed a most prodigious and enchanted place, affording nothing but gusts, stormes and foule weather, which made every navigator to avoid them, as Scylla and Charibidis, or as they would shun the Dieull himselfe, and no man was ever heard to make for the place but as against their willes, they have by stormes and dangerousnesse of the rockes, lying seuen leagues in to the sea, suffered shipwracke."

The first sign of existence on the islands is an inscription on a rock seventy feet above the southern shore, which remained legible for four hundred years before the pollution and acid rain of the latter half of the twentieth century finally dissolved it. It read simply, "1543. . . . . FT," with a crudely chiseled cross. Nothing more is known of the castaway "FT," who sought recognition there in that year.

The first settlement on "the vexed Bermoothes," to use Shakespeare's phrase, came by accident in 1609, when a relief expedition was sent to Jamestown, Virginia. A storm separated the ships before a hurricane caught the flagship, *Sea Venture,* with her captain, Sir George Somers, and 150 men, women, and children on board. Sir George fought the storm well—staying on the poop deck for three days and nights without food or rest, furling all the sails and directing the pumping—but finally gave up. The ship was breaking apart, its hold filled with "9 ffoote of sea water" and the planking ripping loose as tons of water surged back and forth. The men had given up hope, lying down where they could to pray or cry, when suddenly—although they were hundreds of miles at sea —someone cried, "Land, ho!"

And just at that moment, as was recorded in a letter later sent home by the passenger William Strachey, "it pleased God so strangely as the water was staied for that little time." Evidently they had entered the eye of the hurricane, although no one at the time knew anything about such matters. Taking heart at the appearance of land and the cessation of the storm, they sprang to their task, and Somers brought the ship safely to a berth hard upon the outer reef before the storm reasserted itself. They rode it out there and, after it had finally blown away for good, they lowered boats and made it across the sheltered inner lagoon to shore on the Islands of Devils, and found it to be instead a Garden of Eden.

"The Bermooda is the most plentiful place that ever I came to, for ffishe, Hogges and ffowl," Somers later wrote. When he died there, his directions were that his body be shipped back to England for burial but that his heart should be buried on the island. Today the coat of arms of Bermuda features the wreck of the *Sea Venture,* July 28 is still celebrated as Somers Day (although many who do not know the history call it "Summer's Day"), and the local tourist industry is fed by tales of sunken wrecks laden with Spanish gold from the Indies.

6

But hurricanes have done more than play the part of a deus ex machina for the plots of stories and plays. Without them the history of America would have been vastly different.

When Philip II became king of Spain he decided that the two coasts of Florida must be settled and held by Spaniards, to serve as naval bases to protect his gold-laden galleons against French pirates. In 1559 he sent Don Tristán de Luna as captain-general of the fleet and as governor of Florida to find a suitable port on the western coast and to found a colony there.

He succeeded beautifully, at first, by finding a bay sheltered by high land, which would perfectly suit a port city. The city (now called Pensacola) would have been the first permanent settlement in North America, had it not been for a hurricane that blew ashore just one week after de Luna landed and took possession in the name of Spain. Every one of his thirteen ships was sunk, with the exception of a caravel that had been lifted up bodily by the wind and blown inland a hundred yards. (Just a week ago, as I write this in the early spring of 1993, underwater archaeologists reported finding the wreckage of de Luna's ships.)

Forget the west coast, King Philip decided. It was impossible to live in a land tormented by such winds. Instead he ordered de Luna to find a suitable place for settlement on the east coast. And so de Luna put together three ships from the wrecks of his original thirteen and sent them around the tip of Florida toward the east coast.

They never got there. Another hurricane struck and destroyed them all. King Philip decided he could deal with French pirates more easily than with hurricanes, and quickly forgot about Florida and the coast of North America.

7

The French, however, did not. The situation at the time was that the Spanish controlled the seas and so were able to follow up Columbus's discoveries by plundering the Indies. The British were so weak they could do no more than envy the Spanish; the stronger French set out to contest Spanish ownership of the New World.

In 1562 Jean Ribaut led an expedition with 150 soldiers, crossed the Atlantic safely, found the coast of Florida, claimed it for France, and sailed home again. The Spanish, unimpressed, quickly burned his settlement to the ground. Ribaut returned, farther up the coast (in what is now South Carolina), with several hundred soldiers and eighty-one pieces of artillery, and began building a fort that would withstand Spanish assaults. Before he could finish Fort Caroline (which gave the state its name), however, a lookout cried the warning signal as five Spanish warships sailed into the harbor. Having floundered during a hurricane three weeks earlier, they were the sorry remnant of a once-mighty Spanish fleet under Pedro Menéndez de Avilés. The original fleet would have overwhelmed Ribaut, but these few survivors were grossly outnumbered; Ribaut had twelve ships to their five, and quickly he sailed out to give battle.

De Avilés was no fool; he turned and ran, racing down the coast to find shelter in a harbor he named St. Augustine. Ribaut followed, and was about to sail into the harbor and attack when a sudden wind began to blow—not from the sea but from the west, from the land.

They paid no attention, knowing that only winds from the sea gathered enough force to be of consequence. De Avilés gathered his soldiers on deck, preparing to board the enemy ships. His gun ports were lowered, his cannons rolled into place, his sails unfurled to catch the wind and propel them into the fray . . .

Then the rains came, stinging like tiny pebbles. And the rains brought the wind roaring along behind, and the wind brought more rain. Within a few moments the rain had become a solid sheet of slashing water and the wind had filled the sails beyond capacity, splitting them and the masts themselves. Helplessly, men fell overboard and the armada disintegrated into a splintering, foundering horde that disappeared into the swirling mists.

De Avilés, watching from the comparative safety of shore, then made a daring gamble. He gathered his forces together and marched off, by land instead of by sea, to attack the French settlement at Fort Caroline. They marched along what is now Interstate 95, through swamps and across rivers, and found the fort only lightly defended. They captured it easily, and disposed of their prisoners by tying their hands behind their backs and shooting them.

Ribaut's ships were either sunk or broken on the south Florida coast. He gathered his survivors together and he too marched through the swampy forests up the coast to South Carolina, to the hoped-for safety of Fort Caroline. But it was not to be. He arrived there to find the Spaniards in charge. De Avilés greeted him warmly and cut off his head.

De Avilés then attended to business. He was as efficient as he was ruthless, and in the next several years he established settlements along the coast, fortified them, and made allies of the Indians. Had it not been for that hurricane, the Spanish would have lost their foothold in Florida. As it was, the French were forced to spend the next couple of centuries attempting to play catch-up, or attending to the rich fishing pastures off Nova Scotia and Newfoundland. The English, weaker than either of them, had to settle for the barren New England coastline.

And then, because of another hurricane in 1568, the glowering enmity between England and Spain broke out into open hostility, setting the course of the future development of America. The stage was set by the greed of the Spanish rulers and their consequent policy of not allowing their New World colonies to trade with any but their own ships. It was all right for the West Indian colonies to pile up sugar and spices for months at a time, waiting for a Spanish merchant fleet to come and take them home, but they simply couldn't wait all that time for the supplies they needed. And so when an English ship wandered by and offered to take their goods off their hands in return for wine, cloth, flour, and especially fresh slaves from Africa to work the plantations, they usually acquiesced even though it was against the law. Sometimes, to make the illegal trade look good, the governors of the Spanish colonies asked the English to fire off a few cannons so they could tell the Spanish fleet, when it arrived some months later, that they had tried to resist but had been forced into trade with the despicable English.

Despite continual complaints from Spain, English trade in the Caribbean grew and thrived. Among the English sailing baronies was the house of Hawkyns, a Plymouth family who had been in the sailing and commerce business for generations. In 1568 the head of the family was John Hawkyns, and in that year he sailed south and west from Plymouth in the *Jesus of Lubeck* at the head of a convoy of ten ships, one of which, the *Judith,* was commanded by his young cousin Francis Drake.

He coasted along Venezuela, stopping at harbor after harbor, and found nothing but goodwill and fine trading. By July, with the stormy season approaching, he turned for home by way of the east coast of Cuba, intending to head up into the Gulf Stream, which would carry him away from the easterly (west-blowing) trade winds and into the northern westerlies, which would blow him home. But as he reached the Yucatan Channel he was caught from behind by a swelling sea. Those who had arthritis must have felt the twinge in their joints as the atmospheric pressure dropped and the rains came. Before they knew it the winds were upon them.

The hurricane came blowing up from the southeast and blew them northwest, not up along the east coast of Florida but westward into the Gulf. For three or four days it held them in its grasp; Hawkyns and the other captains could barely keep their ships afloat, much less try to maneuver and steer them toward home. When the storm finally left them, the ships were nothing but floating hulks, although grateful indeed to still be floating. They didn't know where they were, but soon they came across a Spanish ship, which surrendered despite the obviously poor condition of the English.

The English needed a safe harbor in which to repair their ships and prepare anew for the long voyage home. The Spanish captain told them there was only one such harbor within easy sailing distance, on an island called San Juan de Ulloa. It was unprotected at the moment, but the Spanish fleet was expected there early in September. Good enough, Hawkyns thought. He sailed there directly, pulled into the harbor, and sent men ashore to assure the settlers he wanted only to buy food and water and repair his ships. Unprotected as they were, they agreed, and the English settled down to work—until the Spanish fleet turned up several weeks early. The viceroy of New Spain hove to offshore at the unexpected and unwelcome sight of English ships in his port. Hawkyns sent men out to explain the situation, swearing they meant no harm and would leave as soon as they were able. The viceroy accepted the situation but also secretly sent word to the Spanish fort at Veracruz, asking for reinforcements.

Several days later they arrived, and the viceroy attacked without warning at dinnertime. The resulting fight lasted well into the night. Hawkyns's ship, the *Jesus of Lubeck,* erupted into flames as a Spanish fire ship came crashing into her. Hawkyns was able to transfer some of his supplies and most of his men into the *Minion,* and slip out of the harbor into open waters. The *Judith,* captained by Francis Drake, had preceded her, but when Hawkyns had time to look around he found he was alone; Drake had sailed for home, leaving him for dead.

The *Minion* was in sad shape; leaks sprung loose by the hurricane were not yet fixed. Half the men on board asked to leave, and once they were well away from San Juan de Ulloa, Hawkyns pulled in to shore (on the coast of what is now Texas) and they disembarked. He then set course for home in the leaky old barge and, miraculously, made it.

Over the next several years others of the expedition came home, one by one, with terrible stories to tell. Twenty-nine had been captured in the fighting at San Juan de Ulloa. Some of the men who disembarked in Texas settled there, but a few years later the Inquisition came to Mexico and spread out along the Gulf Coast. The heretical (Protestant) English were condemned to death, strangled into semi-consciousness, then revived and burned at the stake. And those were the lucky ones; others were sent to live a life of continual torture as galley slaves. Over the next thirty years two of them escaped and made it back to England, where their stories helped to stir up the terrible hatred of the Spanish that shaped the history of the coming century.

Twenty-three of the men who had landed in Texas chose to walk home rather than settle. They headed northeast and, years later, three of them made it to Nova Scotia, where they were picked up by an English ship and taken home. The story they told of a huge, empty, verdant continent waiting to be occupied by good Christian men stirred the imaginations and the hopes of those discontented Englishmen who were praying for a better life, and a place in which to live it.

8

The subsequent history of the United States was formed by those early days when the English, French, and Spanish chose their areas of interest. In the centuries that followed, those who sailed through the Caribbean and Gulf waters began to learn, very slowly, about hurricanes. The storms were frequent, but randomly distributed. You couldn't mount a cruise to study them because you wouldn't have much chance of finding one—and if you did, you'd have a very good chance of being sunk because, aside from being capricious, they were gigantic when measured by all other human experiences.

Initially the Europeans were no less superstitious about hurricanes than the natives to whom they felt superior. When Sir Richard Grenville in 1591 fought the Spanish all night long aboard his *Revenge,* his defeat and death were avenged by the winds of a terrible hurricane that sunk all the Spanish ships. "Over a hundred ships, galleons and merchant ships of the just arrived fleet . . . were wrecked, their crews drowned, their riches lost."

And why? The Spanish understood the reason all too well: "So soone as they had thrown the dead body of the Vice-admiral Sir Richard Greenfield [Grenville] overboard they verily thought that as he had a divelish faith and religion and therefore the divels loved him, so he presently sunke into the bottom of the sea and downe into hell, where he raised up all the divels to the revenge of his death, and that they brought so great storms and torments upon the Spaniards because they only maintained the Catholic and Romish religion."

The English had another explanation: "It may well be thought that it was no other but a just plague purposely sent by God upon the Spaniards and that it might truly be said, the taking of the *Revenge* was justly revenged upon them." But whether it was the "divels" from Hell or God in Heaven who brought the hurricane, all agreed the destruction was total: "For 20 days after the storm they did nothing but fish for dead men that continually came driving on the shore."

. . .

As the centuries passed, sailors began to learn something about the storms that the English variously called *hyrricano, haurachana, uracan,* or *herocano;* that the Spanish called *furicane,* and the French, *ouragan.* The beginning of true knowledge can probably be ascribed to the year 1801, when Colonel James Capper of the East India Company suggested that if one could correlate the personal observations taken by many people at sea who encountered the same hurricane, one might put together from them a comprehensive picture of how the storm is structured, how it moves, and how it develops.

This auspicious beginning was not immediately followed up, but twenty years later an American, William C. Redfield, began publishing the results of his studies, which were a variation on Capper's theme. Redfield had spent time visiting the West Indies and the southern American states, over the previous decade or two, and had encountered several hurricanes and listened to firsthand accounts of others. From these various observations he put together a set of rules that the storms seemed to follow, which upon further contributions from others came to be called The Law of Storms.

All hurricanes, he suggested, form to the east of the Leeward Islands (the chain of West Indian islands beginning with the most northerly of the Lesser Antilles and then stretching southwest from Puerto Rico to Martinique) and then travel a parabolic path to the west, with Bermuda as its focus. They travel at a moderate speed, roughly between five and ten knots, although the winds within them reach speeds upward of one hundred knots and revolve around the center in a counterclockwise direction. Finally, these storms occur randomly, although there is a seasonality to them: they form between June and December, most often between July and October.

But what does all this mean? What causes these storms to form in the eastern Atlantic and then travel in a parabolic path? Moreover, what causes them to revolve counterclockwise, and to form only during certain months of the year? What exactly *is* a hurricane? Aside, of course, from being the curse of the great god Yuracan.

2

OUT OF NOWHERE

1

From *The Miami Herald,* Friday, August 11, 1992:

> Tropical Storm Andrew's wind speeds have dropped to 46 mph
> . . . Its center is poorly organized, its circulation loose and its
> direction away from anyplace it could cause trouble. . . .

No one on August 11 expected Andrew to become a hurricane, yet it
not only did but it turned into a force 5 monster, the strongest class
possible. After becoming a hurricane, it was not expected to pose
any threat to the United States because of its position early on, yet it
not only hit us once but twice, devastating Miami and then going on
to smash into Louisiana.

Why were we so wrong?

2

In order to understand the abnormal, we must first understand the
normal. This is as true of weather as it is of people.

To begin, then, the guiding principle of all air motion, and therefore of all weather, is air pressure: air will always move away from

30

high pressures and flow toward lower pressures. The pressure of air follows directly from atomic and molecular theory.

Air is composed of an assortment of molecules: about 80 percent nitrogen, nearly 20 percent oxygen, and a small variety of others such as argon, water, and carbon dioxide. These are all in random motion, and when any one of them strikes surrounding molecules — or anything else — it exerts a slight pressure as it hits. The pressure from any one molecular collision is infinitesimal, but taking all collisions together it adds up to a lot since there are so many molecules. (There are roughly one hundred times more molecules of air inside a basketball, for example, than there are stars in the entire universe.)

The total pressure of all the molecules in our atmosphere is about 14.7 pounds per square inch. We do not normally feel or sense this because it is exerted in all directions equally. We often make use of air pressure by manipulating this equilibrium. A vacuum cleaner, for example, works by pumping air out of its innards to reduce the air pressure inside. The high-pressured air outside then shoves its way into the low pressure of the cleaner, bringing all sorts of dirt along with it.

To understand hurricanes it is important to know that when a mass of air is heated, molecules move faster. This would increase the pressure it exerts, one might think, since each molecule will hit with more force, which it does; but in the open atmosphere there is another, more important result. As the molecules move faster, they naturally end up farther apart. This makes the air mass less dense than the air surrounding it, and so it will rise (just as a bubble rises in water). A rising mass of air leaves less air on the surface of the earth, causing the pressure to decrease. The normal result is that surrounding air will rush in, just as air rushes into the area of low pressure in a vacuum cleaner, creating winds and moving dirt.

The sun provides the heat that begins this process and that ultimately powers all weather and all winds. In the normal course of events, however, the sun heats the earth nonuniformly, and this varied distribution of heat is what causes the great air masses to move about and create the phenomena that we call *weather* and *winds, climate* and *storms.*

This sounds simple enough, and so it is in theory, but in reality the nonuniformity of the heating varies on scales both great and small, and the resulting complexity is beyond our powers of comprehension.

To compensate, we construct models that approximate the real world and deal with them because they are simpler. This is really the way science and scientists work, in all but the simplest situations. It is preferable, in any science, to deal with a cause-and-effect situation, i.e., to understand the cause of each effect. But, in the real world there are varied causes for each effect, and varied effects for each cause. In the former case, a smoggy day in Los Angeles might be caused by a temperature inversion, which was caused by westerly winds from the Pacific meeting the right combination of northerly weather from Washington, exacerbated by several traffic accidents which tied up traffic and left a half-million cars sitting at blocked intersections belching nitrogen oxides into the air, and where does it stop? In the latter case, an easterly wind over the New Jersey shore not only cools off the coastal resorts but pushes the ambient air somewhere else, which may cause rain over Philadelphia, which blocks the sunshine . . . until it snows in Oregon or hails in Maine as a result.

There are simply too many factors to deal with. This is why a weather report predicting rain tomorrow has as much chance of being right as if the weatherman had picked this forecast by tossing dice (which I think some of them do). The problem is that there are too many variables in the real world: to understand cause-and-effect, you have to isolate one cause and observe what the single effect is going to be, then add another cause and another effect to see how they interact, and keep doing this until you have the entire system under observation.

But this can't be done with the real world. Constructing a model simplifies the world sufficiently so that some things can be understood. Adding complications to the model one at a time begins to approximate the real world. If the approximation is close enough, greater understanding of the world we live in is gained. Unfortunately, we often run into complexities beyond our reasoning capabilities long before the model reaches any true resemblance to our world. Nevertheless, it's the only game in town, so let's play.

The nonuniform heating of the earth that causes all our weather, for example, is a direct consequence of the earth's structure. So let's construct a series of models to imitate but simplify it. To begin with, imagine that the earth is homogeneous—with, say, its entire surface covered with water. No mountains, no land at all, no fog, no ice, no clouds. And further, imagine it to be flat, with the sun shining down on it.

Clearly the surface of the earth would be heated uniformly; that is, each square inch of the earth's surface would receive the same amount of sunlight. (The other side of the earth wouldn't be heated at all, but nobody lives there except "divels" and ignorant savages, so we can ignore it.) The advantage of this model is that it gives us a simple description of the effect of sunlight on the earth; the disadvantage is that it does not bear a very strong resemblance to reality. Making the earth spherical instead of flat adds a complication that brings it closer to the real world. We'll still keep it homogeneous, and ignore the tilt of the earth's axis relative to the position of the sun, so that the sunlight comes down predominantly along the equator all over the world, and we'll ignore the fact that the earth is rotating.

Immediately the situation becomes quite different, for now the unit area of the earth's surface at the equator, indicated by *a* in Figure 2-1, receives all the radiation indicated by the area *A*. But a surface unit of the same size located near the pole *(b)* receives only the radiation indicated by the much smaller area *B*, because of its slanting orientation relative to the sun's rays. Therefore the earth's surface near the equator gets more heat than does the surface nearer the poles, causing the equatorial regions to be hottest.

This is what causes wind. The equatorial heating causes that air to warm and to rise. It doesn't simply rise off the surface and disap-

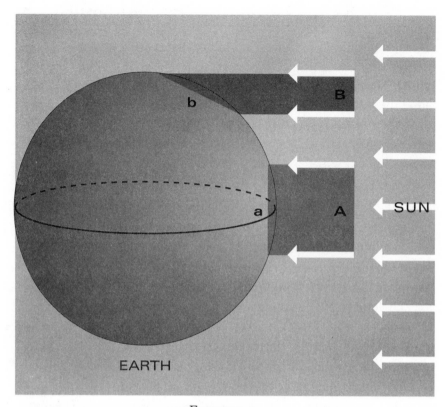

FIGURE 2-1

pear into space, however, because of an atmospheric cap called the tropopause. The air below this cap, the troposphere, comprises what we normally think of as the atmosphere. Rising to a variable height of 10 to 20 kilometers, it is where all the vagaries of weather occur. Within the troposphere, the temperature varies inversely with height: the higher the altitude—so long as you remain in the troposphere—the lower the temperature.

The tropopause marks the point where this regular decrease in temperature stops, and where the stratosphere begins. Here the temperature is constant for a while, then begins to increase again as the ozone layer absorbs the sun's ultraviolet rays.

When warm air rises from the equatorial surface it reaches the tropopause, where it is turned aside. In the northern hemisphere this air mass will head northward, because when it left the surface, it reduced air pressure there, allowing cooler air from the north to spill in. This in turn reduced air pressure up north, and so that is where the rising warm air is pulled when it is shunted aside at the tropopause. As this air mass travels north, it cools off and settles again to the surface, where it is caught up in the general circulation flow and heads back toward the equator. The effect is to establish a circulation cell, as shown in Figure 2-2.

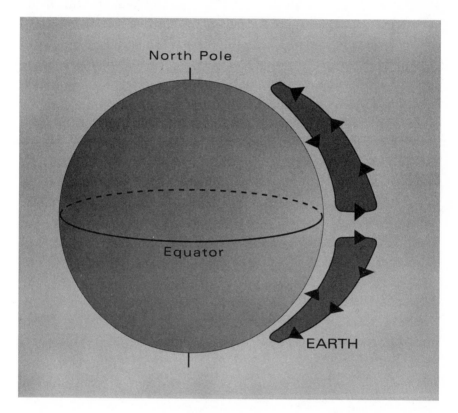

FIGURE 2-2

This general mechanism for air circulation would be accurate if the earth were as we have simplified it: homogeneous and without either a tilted axis or rotation. In this model there are circulation cells for the northern and southern hemispheres. The winds near the tropopause aren't felt on the earth's surface because the rising air doesn't begin to move horizontally (parallel to the surface) until it's up high, near the tropopause, but the cooler winds blowing down from the north are surface winds. So in this model, which begins to approximate the real world, all the (surface) winds in the northern hemisphere are northerly (i.e., blowing from the north).

But of course in the *real* world they're not. This model is insufficiently complex to be useful in understanding our weather/climate system, as it ignores three important parameters: the tilt of the earth's axis, the rotation of the planet, and its heterogeneity.

The tilt is what gives us the different seasons: when the North Pole is tilted toward the sun, the northern hemisphere gets more sunlight than the southern and we have summer in the northern hemisphere, winter in the southern; when the pole is tilted the other way the seasons reverse. While important, this is not partic-

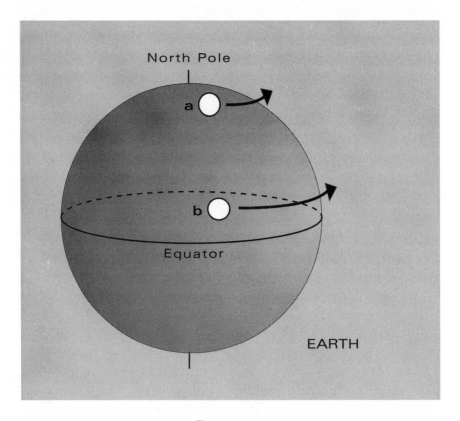

FIGURE 2-3

ularly relevant to the questions covered here. The rotation, however, is.

Consider two parcels of air in our circulation cell—one at the surface near the pole and one up high but near the equator—and let's look at them from a different angle (see Figure 2-3).

Both air masses are rotating along with the rest of the earth, from east to west. But it's important to note that the velocity of *a* is less than that of *b* (as is the velocity of the earth right under them): both parcels, and the earth over which they float, are moving to the east at a speed that will take twenty-four hours to make one entire revolution. But the distance to be covered by the equatorial parcel is obviously much greater than that to be covered by the one near the pole, since *a*, very close to the pole, will have to move only a few miles to complete its rotation, while *b*, at the equator, will have to move around the entire circumference of the earth. Therefore *b* is moving much faster than *a*, as indicated by the arrows.

This has to affect the circulation cells. Parcel *a*, at the earth's surface, is moving south, while parcel *b*, up high, is moving north. But as they move along the cell they still retain their original easterly motion (conforming to Newton's law, which states that a particle in motion tends to remain in motion in its original direction unless some force acts to change it). So added to their southerly and northerly velocities is an easterly velocity, but with a difference.

As *a* moves south, it is moving over earth that is also rotating toward the east, *but at a faster speed;* as *b* moves north it is moving over earth that is rotating east at a slower speed. So *a* sees the planet beneath it moving east faster than it can follow; in other words, relative to the earth's surface, *a* begins to drift westward. For *b* the opposite happens: it moves eastward across the face of the planet (see Figure 2-4).

Thus one portion of the circulation cell moves east and another moves west, creating divergent motions that tear the cell apart. The turning of the north-south winds to east and west is known as the Coriolis force, but it is an apparent force only, due entirely to

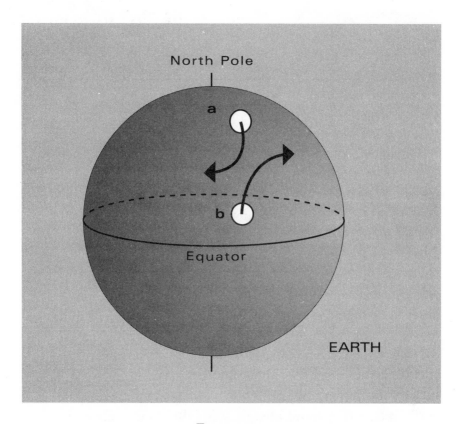

FIGURE 2-4

the rotation of the planet. The result of the Coriolis force is that the original cell breaks apart into three smaller cells, in each of which the differences in easterly velocity are small enough to be contained. The two southern cells are named for people who did the original work describing them: a Hadley cell rises over the equator and downdrafts at about latitude 30 degrees north, and a Ferrel cell downdrafts next to the Hadley and rises at about 60 degrees. Finally, there is a Polar cell circulating from the Ferrel cell to the North Pole (see Figure 2-5).

There are several consequences to this model of the atmosphere. First of all, note the direction of the circulation: where the cells touch, their direction of air flow is the same. This means, for example, that at about latitude 30 degrees north there is a convergence of downdrafting cold air from up high toward the surface, causing what is known as the subtropical high pressure area,

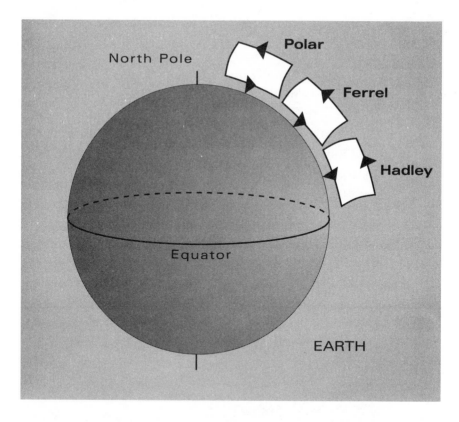

FIGURE 2-5

while at about 60 degrees there is an analogous updraft causing a polar low.

Let's follow the course of the subtropical high in more detail. The motion of any air mass relative to the planet's surface is called *wind,* and so this cold air mass plummeting toward the surface is going to become a rather steady source of wind. The descending air diverges at the surface, flowing both northward and southward from latitude 30 degrees. We began this discussion to talk about hurricanes, and since these form in the region below 30 degrees, we'll follow these latter winds. Flowing southward, they turn to the west because of the earth's rotation, and thus become a prevalent wind from east to west: the easterly trade winds.

These are the winds Columbus found when he sailed west from Spain; he hadn't known about their existence, but without them the journey to the Indies would have taken him so long that he and all his men would have died of scurvy long before they were able to freshen their supplies. Initially the crew gave thanks to God for the steady winds; later, as weeks passed and the winds persisted, they grew worried that they would never get back to Spain beating against those winds. Luckily, they discovered that if they sailed north from the Indies instead of trying to force their way east, they could escape them. The same circulation cells that form easterly winds below 30 degrees create westerly winds above it, as the northward flowing air beneath the Ferrel cell diverges to the east (see Figure 2-6).

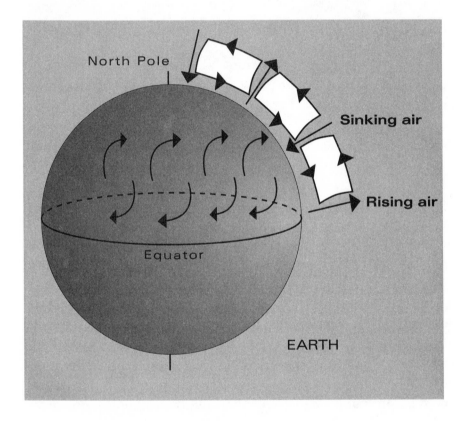

FIGURE 2-6

As the southward-flowing air reaches the equatorial region it be-gins to rise again, lowering air pressure, and the winds die down. This entire circulation pattern is duplicated in the southern hemisphere, of course, and so immediately over the equator we have a region of rising warm air and a near total absence of sustained winds. Sailors who ventured into this area soon became becalmed, and many spent their last days watching their supplies of drinking water dwindle and food turn rotten, and waiting in vain for a wind that never came. This area is called the doldrums, giving rise to the nonnautical use of the word for when we are dull, listless, and depressed. (Or maybe vice versa; no one seems to know for sure.)

A similar region of wind loss occurs at about latitude 30 degrees, where the converging high-altitude air drives downward, before it spreads north and south to become the trade winds. This air is clear and dry; since it has been long separated from the surface and has lost most of its moisture, it often leads to desert areas where it comes down over the continents. Over the oceans it gives clear skies, without coherent lateral movement and therefore without prevailing winds. These areas are known as the horse latitudes because crews of ships becalmed there would begin to eat the horses on board after their own stored food spoiled.

So the model now is quite a reasonable approximation of the real world: it accounts for the trade winds, the doldrums, and the horse latitudes. It is indeed a sufficient approximation, since between the doldrums and the horse latitudes is where hurricanes form.

3

Sailing ships are not the only objects pushed around by the winds; air masses in general, and storms in particular, will also be guided by them. But before we can understand storms, we have to lay the groundwork for the most obvious characteristic of a hurricane—its counterclockwise rotation. This is usually explained in terms of the Coriolis force, but it is more helpful, I think, to view it in terms of the conservation of angular momentum.

There are a whole host of conservation laws in this universe, and they constitute one of the most useful tools we have for interpreting nature, once we accept both the principle that they exist and the fact that we don't understand this principle at all. The laws themselves are simple: some things in the universe are *conserved,* neither created nor destroyed in an isolated system — that is, a system which doesn't interact with the rest of the universe. Consider, for example, the room you're sitting in, and the principle of conservation of mass. Mass is not going to appear or disappear in this room. The sofa is not going to evaporate into nothingness, nor will clumps of mass — nor even a single atom — suddenly appear where none was before. If the system is not isolated, of course, such things can happen. Another person can walk in the door, and thereby introduce more mass into the room; but this is not a creation of mass in isolation, since the entering person represents a connection with a larger system, the universe.

This conservation of mass is so logical that we take it for granted, but in fact there is no a priori reason for it. One could visualize a universe in which this law of conservation of mass simply did not exist. It would be a chaotic universe, one in which it would be impossible to plan for the future — your house or spouse or bank account might disappear without warning — but there is no reason it could not exist. We simply have to accept the fact that we happen to live in a universe where mass is conserved, although we have no idea why. (Possibly there *is* no "why"; it simply happens to be so. Or perhaps there is a "why" that we don't yet understand. And of course this "law" of mass conservation isn't strictly true: Einstein demonstrated that $E=mc^2$, which says that mass can be converted to and from energy, in which case the mass can appear or disappear so long as there is a concomitant disappearance or appearance of energy. Strictly speaking, the law is one of conservation of mass-energy.)

Aside from the obvious conservation of mass and energy, there is a variety of other less obvious laws — such as conservation of parity, baryon number, and angular momentum. It's the last of these that we have to consider. Angular momentum arises when a mass moves along a curved path, as when a stone tied to a string is whirled in a circle. The stone's angular momentum is defined as its mass multiplied by its velocity and by the length of the string; in general, angular momentum is given by $M \times V \times D$, where D is the distance from the mass M to the center around which it is revolving with velocity V.

There is no immediate reason why we should define angular momentum in this manner, other than the fact that repeated observations have shown the quantity thus defined to be conserved. It's important here since it explains why hurricanes rotate the way they do, and because this rotation is one of the secrets to understanding exactly what a hurricane is and when and how it forms.

Imagine a single mass of air extending from the pole to the equator, not moving relative to the earth (which means no wind is blowing) (see Figure 2-7).

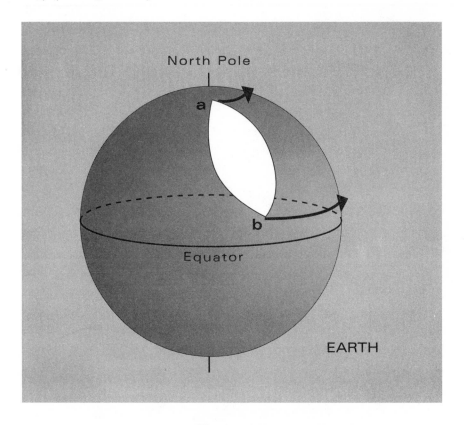

FIGURE 2-7

Admittedly this picture is not very tightly bound to reality—as we have seen, such a mass of air would break up into smaller masses very quickly—but bear with me a moment. The air is stationary with respect to the earth beneath it, but since the earth is rotating west to east, so is the mass of air, as shown by the arrows. Consider the two end portions of the mass of air, labeled *a* and *b*. As noted before, they are both rotating eastward along with the rest of the earth, and at two different speeds, with *b* moving faster than *a*. This means that the air mass, although stationary over the face of the earth, is actually rotating. To see this better, replace it with a small rod, the one to the left in Figure 2-8, with the length of the arrows indicating the magnitude of the velocities of each end of the rod:

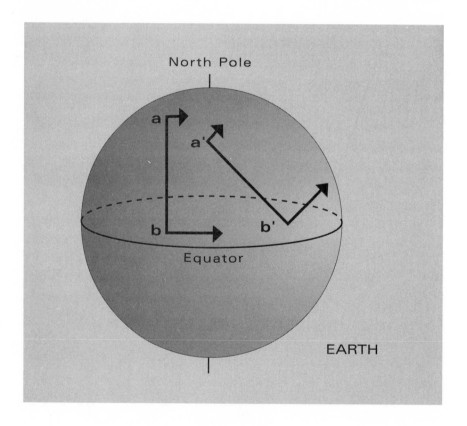

FIGURE 2-8

If the lower end is moving faster than the upper end, the rod will have to rotate, as shown in the figure to the right. And because of the direction of the earth's rotation, the more equatorial part of the rod—or the air mass—will always be moving faster toward the east, leading to a counterclockwise rotation, as shown.

So we have a counterclockwise-rotating air mass, even though it is not moving relative to the surface of the earth (which means, remember, that no wind is blowing at the moment). Suppose that heat is applied to the center of this air mass, so that the air there begins to rise. This creates a partial vacuum near the earth's surface—or at least an air mass with lower pressure—so the rest of the air mass, at *a* and *b,* will move toward the center, resulting in a contracting, rotating mass of air. Since the total mass of air *(M)* remains the same, and the distance *(D)* decreases—where *D* is the distance from the ends of the air mass to its center—in order to conserve the total angular momentum the velocity *(V)* will have to increase. In other words, this mass of air will now begin to spin faster. Since the earth doesn't spin any faster, the air mass now begins to move relative to the earth's surface, and winds begin to blow counterclockwise around its center.

This is the same principle employed by ice skaters. A skater enters a spin with arms flung wide, creating a certain amount of angular momentum. Without any further effort, she can increase her rate of spin simply by bringing her arms in and crossing them over her chest: she has decreased the distance of part of her body from the center of her rotation, and so without any expenditure of energy, her spin will speed up. To slow down and stop she simply flings her arms wide again, increasing *D* and therefore decreasing *V.*

It is not necessary that the air mass reach from the North Pole to the equator; I just drew it that way to make it easier to see the effect. Any mass of air will rotate similarly, with the southern end moving faster than the northern, therefore spinning counterclockwise as in the example. (This is for the northern hemisphere; in the southern hemisphere the air spins clockwise.) As the warm air leaves the central portion of the mass, the outer parts contract to fill the void, and as they contract they spin faster and faster in order to conserve angular momentum, and away we go: we've got counterclockwise winds blowing, and as the air mass contracts, the spin speeds up and the winds blow harder and eventually we've got a hurricane.

This counterclockwise rotation was discovered through observation long before the explanation for it was found. In 1821 a hurricane struck New England, where a young man named William Redfield was in the saddling business. A large part of his time was spent walking through the countryside selling his services as a harness maker. As he walked north through the eastern parts of Connecticut and Massachusetts a few weeks after the storm, he noticed that all the trees blown down lay with their limbs pointed to the northwest, and as he circled around at the end of his journey and walked home through the western parts of those two states he realized that the downed trees were pointed toward the southeast. Curious.

He could not have been the only one who looked at the wreckage of that storm, but he was the only one who noted that pertinent fact. He began to collect records of the storm, and in the years that followed he found statements from people in Baltimore and Virginia, Delaware and Pennsylvania, New York and Connecticut and Massachusetts, and he put them all together and correlated them geographically, and realized that the pattern was not that of a unidirectional wind, but rather of one that spun counterclockwise.

Ten years later while working on a boat in the Hudson River he got into conversation with another passenger who was also interested in hurricanes. This was Denison Olmstead, a professor at Yale University, who invited Redfield to write up his observations for the *American Journal of Science.*

At that time no serious scientist bothered to read the journal. In science, everyone who was anyone was European; American scientists were regarded as coonskin-wearing frontiersmen or cowboys, fit only for the entertainment of young boys. But in 1831 an Englishman who was only sort of a scientist, Colonel William Reid of the Royal Engineers, read Redfield's article and was intrigued by it. A couple of years later, when he was ordered to Barbados to re-build the structures damaged by a hurricane, he remembered it and, intrigued by the awesome damage he observed on the island, he determined to take Redfield's ideas a step further. He searched the historical records and came up with the idea of plotting the characteristics of The Great Hurricane of 1780. As an early account puts it:

> The most terrible cyclone of modern times is probably that of the 10th of October, 1780, which has been specially named the "Great Hurricane." Starting from Barbados, where neither trees nor dwelling were left standing, it caused an English fleet anchored off St. Lucia to disappear, and completely ravaged this island, where 6,000 persons were crushed under the ruins. After this the whirlwind, tending toward Martinique, en-veloped a convoy of French transports, and sunk more than 40 ships carrying 4,000 soldiers; on land the towns of St. Pierre and other places were completely razed by the wind, and 9,000 per-sons perished there. . . . Most of the vessels which were on the path of the cyclone foundered with all their crews. Beyond Porto Rico the tempest bent to the northeast, toward the Ber-mudas, and though its violence had gradually diminished it sunk several English warships returning to Europe. At Bar-bados, where the cyclone had commenced its terrible spiral, the wind was unchained with such fury that the inhabitants hiding in their cellars did not hear their houses falling above their heads.

Reid picked that particular storm as his data source not because of its destruction of Barbados, where he first saw a hurricane's ef-fects, but because of a few phrases in the description above: *an En-glish fleet anchored off St. Lucia . . . a convoy of French transports . . . several English warships.* They indicated what for Reid was a bit of luck: England and France were at war in 1780, and so the Atlan-tic was dotted with warships that doubtless kept a daily log of ev-erything that happened. Furthermore, because of the progress of the war, most of these ships were concentrated around the Indies, where the hurricane roared right through. To add to the peculiar circumstances, since the ships were at war they did not all flee the approaching hurricane; many were ordered to remain in harm's way for military reasons. (The admiral in charge of the British ships unsurprisingly stayed snug and safe in New York Harbor.) Finally, the military logs (unlike those for commercial or private ships) were eventually placed among the public records in London.

Fifty years after the hurricane struck, Colonel Reid went through all these documents at the British Public Records Office. From statements such as "There arose a hurricane at NE which increased by the morning to a degree of violence that is not to be described . . ." and from the known positions of each ship recording the event, he put together a picture of exactly how the winds blew. Thus, since 1831 everyone has known that the winds of a hurricane blow counterclockwise. With this information it became possible for ships at sea to discern the location of distant hurricanes, and so to avoid them. Previously, a ship's captain feeling a strengthening wind in his face would naturally assume the storm was in front of him. But now he would know that if the wind was in his face, the hurricane generating it must lie to his right; if the wind came from his stern, the storm must be to his left.

In 1944 Ivan Ray Tannehill, chief of the Synoptic Reports and Forecasts division of the U.S. Weather Bureau, published the first modern scientific history of hurricanes. In it he wrote: "A knowledge of the law of storms is an essential part of the education of ships' officers. To the landsman who experiences a tropical storm, the direction from which the wind blows, in relation to the position of the storm center, is sometimes puzzling. . . ." But to the educated ships' officer, it was as plain as the nose on his face. "The student navigator is told how to judge the location of the storm center, how to maneuver his ship to avoid it, and how to anticipate changes in the progressive movement of the storm so that he may alter his course accordingly." By 1944, 113 years after Reid's work, everyone knew how to avoid a hurricane at sea. Everyone, that is, except an admiral or two.

In the late afternoon of December 17, 1944 (just months after Tannehill's book was published), the United States was at war with Japan. Task Force 38 was sailing eastward of the Philippines, arranging a rendezvous with its fueling group in anticipation of their assignment supporting the coming invasion of Mindoro. For the past few days exceptionally strong winds had been blowing out of the northeast, growing in strength each day, and by the eighteenth the flag officer of Task Force 38, Admiral William F. "Bull" Halsey, who was on his way to becoming probably the most famous naval officer of World War II, began to worry about whether the seas might be too rough for the refueling operation. He turned to his weather officer and asked what the hell was happening.

The aerologist reported that a storm seemed to be blowing up somewhere to the east and moving west (toward them), but that a cold front was expected to intervene and blow it back out to the northeast. Okay, Halsey said, and ordered a new rendezvous point farther to the west.

As dusk was falling he was handed another weather report. A scout plane had reported sighting the storm at 0500 that morning; it was much closer to the fleet than the weather officer had suggested.

Halsey blinked and looked at the report again. *At 0500 this morning?* Why the hell had it taken twelve hours for the report to reach him? Because of the usual things that happen in wartime, as it turned out. First, the pilot of the scout plane had been under orders to maintain radio silence unless something very important was sighted, and he had been given to understand that the phrase "very important" meant battleships and aircraft carriers; he didn't think a mere storm met the criterion. Second, when he returned to his ship and handed in the report, it was encoded before being sent on to the admiral. This was all very well, except that the encoding officer also didn't think it very important, and so he used the usual weather code instead of the military machine-encrypted code, and the usual weather code was both encrypted and decoded by hand, and the process was very slow. By the time Halsey got the weather report, the storm had had some twelve hours in which to move, and he could only guess at its current position. As it turned out, he guessed wrong.

The exact details of what transpired do not remain; but Halsey had to guess either that the storm represented only a greater-than-normal blow or presaged a hurricane (or, as they call it in the Pacific, a typhoon). If it was just a big blow, the modern warships could ride it out. If it was a typhoon rising, they had better run. Evidently Halsey decided that it looked like a typhoon, for Task Force 38 ran from it. But they ran the wrong way.

They headed west to flee, but proceeded into increasing winds and churning seas. It became too rough for the refueling operation; Halsey had to find calmer waters, and this is where he made a tragic mistake.

Since a typhoon is nothing but a hurricane by another name, its winds have to be blowing counterclockwise. If they were blowing from the northeast, the center of the typhoon had to lie to the southeast. (Facing southwest, with the wind on his back, the storm center lay to his left.) But Halsey evidently thought the storm was to the northwest, for he ordered Task Force 38 to steam full speed to the southeast, directly at the approaching monster.

How could he have made such a mistake? Perhaps because of the confusion surrounding the interpretation of the Coriolis force. This force induces winds pouring out of a pressure high to turn clockwise, and forces winds pouring into a pressure low to turn counterclockwise. Perhaps Halsey's weatherman didn't know that a typhoon center is a pressure low; more likely it was Halsey who didn't know, and didn't bother to ask. At any rate the task force steamed southeast, right into the full force of the typhoon.

Task Force 38 was low on fuel, and as a consequence its ships were riding high in the water. The smaller ships, the destroyers, upon encountering rough weather would normally fill their empty fuel tanks with seawater as ballast, but if they did that now they wouldn't be able to empty them again and refuel in time to support the Mindoro landing. Instead, they decided to try to refuel in the stormy seas. Within the hour they were forced to stop; the weather was growing worse by the minute and the operation was impossible. Halsey radioed MacArthur that he wouldn't be able to support the invasion. Ordinarily he would have worried about MacArthur's reply; today he had other things to worry about, such as men swept overboard from the tempest-tossed destroyers. In that growing horror of a sea there was nothing he could do to rescue them, nothing he could do except try to ride it out. Worse reports soon came in: airplanes were being swept overboard from the flight decks of his carriers; below decks, planes were breaking their moorings and crashing against the walls, bursting into flame; high-octane fuel was spilling over the floors; five-hundred-pound bombs and two-thousand-pound torpedoes were breaking loose and rolling about; .50-caliber machine-gun bullets were scattered amid the growing fires. And the raging typhoon rose in intensity, and rose again.

"No one who has not been through a typhoon," Halsey later wrote, "can conceive its fury. The 70-foot seas smash you from all sides. The rain and the scud are blinding; they drive you flat-out until you can't tell the ocean from the air. At broad noon I couldn't see the bow of my ship, 350 feet from the bridge. The *New Jersey* once was hit by a 5-inch shell without my feeling the impact; the *Missouri,* her sister, had a kamikaze crash on her main deck and repaired the only damage with a paintbrush; yet this typhoon tossed our enormous ship as if she were a canoe."

The *New Jersey* was a battleship. There were destroyers out there in that raging wet inferno, small destroyers, unstable with their low fuel load. Nearly eight hundred men died. The destroyers were rolled over so severely that their smokestacks dipped into the water, which then poured inside. Three sank without a trace. Out of 831 men on board the three ships, only seventy-four were later rescued. Another thirty men were swept overboard from the larger ships. A hundred and eighty-six airplanes were lost. Instead of supporting the invasion, Task Force 38 went home and licked its wounds, thankful to make it back to port.

Fleet Admiral Nimitz was not thankful; he was pissed. As Halsey's boss he was the one who had to explain to MacArthur why the assigned task force had not shown up. He convened a court of inquiry, which duly reported that the tragedy was Halsey's fault, although the final report contains the words: "[The] errors of judgment [were] committed under stress of war operations." Nevertheless, it had been more than a hundred years since Colonel Reid had spelled out a hurricane's circulation, and it was inexcusable for a man in charge of other men's lives to have remained ignorant of the seas he was sailing. As Tannehill had written, "The student navigator is told how to judge the location of the storm center . . ."

Nimitz wrote a letter of advice to Halsey: "The time for taking all measures for a ship's safety is while still able to do so. Nothing is more dangerous than for a seaman to be grudging in taking precautions lest they turn out to have been unnecessary." Six months later, off the coast of Okinawa with another task force, Halsey once again spotted the oncoming winds of a typhoon. And once again he sailed directly into it. (They didn't call him "Bull" for nothing.) This time he was luckier; he lost only a half-dozen men who were swept overboard.

Plus ça change, plus c'est la même chose.

Well, at least now we understand not only that a hurricane has to turn counterclockwise, but also why it does so. All that remains to be explained is how the central portion gets heated, and what happens to the warm air that is escaping upward.

4

Again, the sun is responsible. A hurricane is basically a huge heat pump that gathers the sun's heat from a large area and pumps it into a concentrated region, warming the air and making it rise, sucking in air from the outskirts to fill the void, forcing it to rotate faster and faster as it continually collapses in on itself.

But the sun doesn't simply work on the burgeoning system, heating up a large mass of air and starting a storm. The process is much more complicated, more subtle, than that.

One of the few science-fiction stories I remember from the long-ago days, when I read *Astounding* and *Fantasy, Galaxy* and *If,* dealt with time travel, which was a tourist business then (sometime in the far future), and the most important rule was to never step off the designated walkways during a trip into the past. Doing so meant interacting with that world, and any interaction in the past could conceivably change the world of the future in vast and unknown ways.

The particular time traveler who was the protagonist of the story, while walking along the designated pathway through a primeval forest, saw something—a flower I believe—that he wanted to observe more closely. No one was looking, so he stepped off the pathway to smell the roses, metaphorically, and in doing so he inadvertently crushed a caterpillar, which then did not grow up to be a butterfly, which then did not spread a bit of pollen, which then did not create a plant, which then . . .

When the time traveler returned to his world a million or so years later, he found it incomprehensibly changed. Everything was the same, yet different. The small change, the absence of one butterfly, had been magnified by each succeeding generation over the millenia, resulting in a world that was not the same as the world he had left, a world subtly but terribly and awfully different.

I remember that story, although I have forgotten nearly every other one I read in those days, because it is true—at least with regard to weather, climate, and hurricanes. The interactions of our atmosphere are so complex, and reach over so wide a geographic range, that the results of any small change are incomprehensible and may well be gargantuan. Meteorologists, in fact, are fond of saying that a hurricane over the Atlantic can begin with a butterfly's beating wings in Vietnam.

Rather than trying to follow the story all the way from the butterfly, let's accept the fact that variations in the atmosphere occur continually and seemingly at random. (They're not *really* at random, but they are due to causes too small to observe, like the beating of the butterfly's wings.) Imagine a river lazily rolling along: swirls and eddies appear and disappear continuously; so it is with the river of air that flows all around us. We feel these swirls as winds, and we measure the eddies as regions of high or low pressure. A particularly complex combination of swirls and eddies is necessary for a hurricane to develop. Some of these circumstances are relatively normal, others are more and more unlikely; all of them must come together if the hurricane is to be born.

To begin with, the normal circulation of air under the northern hemisphere's Hadley cell leads to the steady northeasterly surface winds that we call the trade winds, blowing from about latitude 30 degrees down toward the equator and curving out westward across the Atlantic Ocean. In the southern hemisphere the same convection pattern leads to southeasterly winds, and at the equator these two bands converge. The convergence of surface winds in this region leads to its name: the intertropical convergence zone. Occasionally an elongated area of low pressure forms within this zone; when it does, the prevailing winds will push it along to the east, and it is known as an easterly wave (see Figure 2-9).

This is circumstance number one: an occasional but normal swirling of the atmosphere leads to a low-pressure easterly wave. Circumstance number two is more unusual: the intertropical convergence zone must be displaced from the equator if an easterly wave is to become the focus point for a growing hurricane. This can happen, but only occasionally. Why is it necessary? Because hurricanes cannot develop on the equator itself. To understand this, visualize a mass of air centered directly over the equator. Its most northerly and most southerly points will have

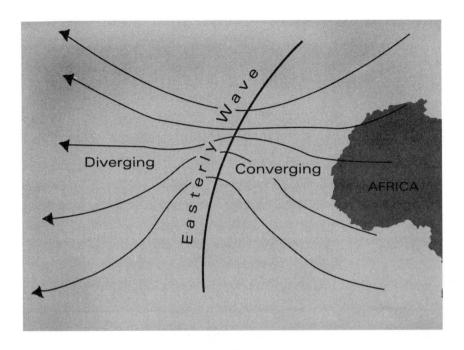

FIGURE 2-9

an identical eastward velocity, and so its contraction will not lead to any spinning motion. The contracting air must be displaced north or south from the equator to get the differential velocity needed to start the spin. If it is only slightly off the equator, there is not sufficient difference in velocities to initiate a hurricane's necessary spinning; it has to get at least 7 degrees away.

And, as we said, this does occasionally happen just from the random swirling of the atmosphere. (Remember, not really random; but whatever causes it is too small a perturbation to be recognized.) Take a look at the easterly wave pictured in Figure 2-9: on the western side the winds are diverging, but on the eastern side the normal convergence is intensified. If the winds are strong enough and the wave is deep enough, the convergence is sufficiently strong to initiate the spinning process. The converging winds build up air pressure on the easterly side of the wave—which should strike us as odd since we know that the center of a hurricane is an area of *low* pressure. Something must be missing from this picture.

What's missing is the heat of the tropical ocean. As winds cross over the ocean and converge on the easterly wave, they pick up heat radiated from the ocean. Thus winds approaching the wave are getting progressively warmer, so the converging winds are warmer than those pushing in from farther away. This lighter, warmer air will necessarily rise and maintain low pressure within the wave. In essence, the rising winds suck in the converging winds, and the storm system begins to grow. This progressive heating of the converging winds is a necessary condition; without it the convergence would increase pressure at the wave and destroy it. The system would disperse and disappear instead of intensifying. This explains why hurricanes form only over the tropical oceans, and only in the warmest months of the year.

To continue, the winds pouring into the easterly wave have picked up not only heat but also a lot of moisture from the ocean below. This is explained by a peculiar combination of physical laws: the warmer a mass of water gets, the less gas it can hold; while the warmer a mass of gas gets, the more water it can hold. Actually, it's not all that peculiar: both facts follow directly from the molecular structure of things. A mass of water is actually a bunch of loosely bound H_2O molecules wandering around rather freely, in constant motion. In water's solid form, ice, each of the molecules is locked in place. But when heat is added the molecules begin to vibrate faster and faster until, at the melting point, the vibrations break the molecules loose from their assigned positions.

If the liquid water is heated further, the molecules will move even faster. What happens at the surface? A molecule rushing upward might have enough energy to break loose from the intermolecular pull of all the other water molecules and dart right out of the water's surface and into the air above: it has become a molecule of water vapor, or gas. As water is heated each molecule gains in energy and speed, resulting in more molecules hitting the surface per second and a greater chance that any given molecule will have enough energy to break free. Simply, as water gets warmer, more of it evaporates.

The opposite situation obtains within the air mass above the water. Imagine a bunch of liberated water molecules floating around up there. They remain isolated in the gas or vapor phase; if they coalesce and form bunches of molecules, and eventually become big enough to form a drop of liquid water, gravity will pull it back down. At this point the water has reverted to its liquid state and begins to precipitate out of the air as rain.

If a mass of air is hot, all its molecules—including the H_2O molecules—race around full of energy. If two of the water molecules collide, they are likely to bounce off each other and continue on their separate ways. If the air mass begins to cool off, however, the molecules lose energy and move more slowly; colliding water molecules are now more likely to stick together (the weak intermolecular binding force will have a bigger chance of grabbing hold of them). So as an air mass cools, it loses water; conversely, as it warms, it can hold more of it as vapor.

Clearly, the converging winds that blow across the warm tropical oceans collect H_2O molecules breaking the surface of the water; by the time the winds reach the easterly wave they have absorbed a lot of water. This absorption is absolutely necessary because the water vapor becomes the fuel to drive the hurricane's engine. Now the winds pile into the easterly wave, building in pressure, and as there is nowhere to go but up, the masses of converging air begin to rise. As they do, they get cooler, which is rather a peculiar thing when you think about it.

Of course, we never do; we just take it for granted that air gets cooler at higher altitudes. But why should it? If the sun is the source of all our warmth, shouldn't the air get warmer if we go higher?

To answer this touches on some of the biggest questions we face today: why is the earth's atmosphere so warm, and is it getting warmer? Clearly, what we are talking about here is the greenhouse effect.

5

The sun emits most of its energy as visible light. This traverses the nearly 100 million miles of space that separates sun from earth, and passes through the atmosphere without interacting with it in any significant way. Ours would not be a particularly pleasant planet on which to live if sunlight interacted with the atmosphere—that is, if it was absorbed by the air, or if it bounced off the atmosphere and back into space—as we would live in a dark world, unable to see.

(Well, to be absolutely honest, the sunlight does interact with some particles in the atmosphere, scattering the red portion of the spectrum somewhat more than the blue. This is what lets the blue light come through to us, so that the sky above appears to be this color. And when the sun is low on the horizon, this scattered red light, which would otherwise be lost to our sight, is scattered back down to us, giving us lovely sunrises and sunsets. But most of the sun's light comes right through to us.)

If light were to pass straight through the atmosphere without interacting, it would not warm the atmosphere; "warming something" means passing energy to that something. But when the sunlight reaches the surface of the earth, it does interact. Clearly it does, for it doesn't pass right through the earth as it does through the atmosphere. Some of the sunlight is simply reflected back into space—by snow, for example—but much of it is absorbed. And since light is a form of energy, this energy is passed on to the earth, which then warms up.

If the earth existed without the sun, its surface temperature would approach absolute zero (−273 degrees Centigrade) since its internal heat is hardly felt on the surface. Very simply, we are warm because there is a sun and its light penetrates the atmosphere, is absorbed by the earth, and warms it.

But when a body gets warmer than its surroundings, it begins to emit heat in the form of electromagnetic radiation, in accordance with the Second Law of Thermodynamics. So as soon as the earth is warmed beyond the temperature of the surrounding space (which is essentially absolute zero), it begins to emit its own heat radiation. Obviously it doesn't emit as much heat as the sun does; less obviously, it also doesn't emit it at the same wavelength, for the hotter a body is, the more heat energy it emits and the shorter the wavelength of its radiation. The sun emits most of its energy in the region of visible light; the earth emits far less energy at a much longer wavelength, which our eyes cannot see. This is the reason we don't see the earth glowing beneath our feet. Nonetheless, it is emitting radiation all the same, in the region of the spectrum we call infrared.

Infrared radiation does not pass through the atmosphere freely, as does visible light. Different molecular structures behave differently with regard to the passage of electromagnetic radiation through them. The relation is a complicated one, depending on the wavelength of the radiation and the size, structure, and intermolecular distances of the structure, but the end result is that some substances allow radiation of certain wavelengths to pass through while obstructing other wavelengths. Our skin, for example, allows X rays (short wavelengths) and infrared (long wavelengths) to pass through while denying passage to visible light, which has wavelengths between these two extremes. The result is the usefulness of X rays to look inside our bodies, and the use of infrared lamps to provide "deep heat" to our muscles.

Glass is another common substance that differentiates between radiation of different wavelengths. It allows visible light to pass through, but restricts the passage of ultraviolet and infrared. Ultraviolet light, with wavelengths just a bit shorter than visible, comes to us from the sun in nearly as great intensity as visible, and does all the harm we hear about: it tans us but also burns us, causes skin cancers and cataracts. All this even though it is largely absorbed when it tries to pass through the ozone layer at the base of the stratosphere, through which the visible light passes freely; without this protection, terrestrial life would be impossible. The ultraviolet also doesn't pass through glass, which is why you don't get a sunburn sitting in a car even though the sun is shining through the windows.

Glass is also largely impenetrable to infrared, which is why a car in the sunshine on a summer's day heats up like an oven inside. The visible light, carrying the sun's energy, passes through the windows and is absorbed by the material inside: the seat, steering wheel, floor, etc. This material, as it warms up, begins to emit heat energy in the form of infrared light, which cannot pass easily through the windows. With a large influx of energy—visible light from the sun—and a small outflow of energy, the car interior gets hotter and hotter, until so much infrared is being emitted that its slight seepage through the windows will balance the incoming radiation.

Called the greenhouse effect, this is the process responsible for keeping glasshouses—greenhouses—warm even during a cold winter in Minnesota. It's important on a global scale because certain molecules in the atmosphere create this same effect. Water vapor, carbon dioxide, methane, and chlorofluorocarbons all allow the free passage of visible light while trapping the infrared. Because of this, the atmosphere is not directly warmed from the sun above, but indirectly from the earth below. The sun's light passes through without effect and warms the earth; the earth then radiates infrared light, which is absorbed by greenhouse gases in the atmosphere, and this absorption of energy provides warmth to the air. Without this absorbing atmospheric blanket the sun would not be able to warm the earth above zero degrees Centigrade. All the oceans would be frozen solid, and life would never have evolved here.

We see clear evidence of the efficacy of this effect if we look at the earth's closest neighbors in the solar system. Although Venus is closer to the sun than the earth, and Mars is farther away, the differences in the three relative temperatures due to distance are small: Venus should be a few tens of degrees warmer than us, Mars should be slightly cooler. But in fact Venus has a surface temperature near 500 degrees Centigrade, while Mars is well below freezing. This is because Venus has a thick atmosphere loaded with water vapor and carbon dioxide—what is called a "runaway" greenhouse effect—while Mars has only a thin atmosphere with virtually no carbon dioxide or water. This is what scares most scientists today: we are increasing our own levels of atmospheric carbon dioxide, and Venus offers a frightening picture of what that could mean. But more of that later.

As far as we are concerned right now, it makes sense that as you go higher in the sky—farther away from the warm earth—the air gets colder. And so as the warm winds, loaded with water vapor from their long journey over the tropical oceans, lump together into the easterly wave and begin to rise, they also begin to cool. Doing so, the air loses its ability to retain water vapor. And when that happens, heat is released in accordance with the inescapable law of conservation of energy.

6

Conservation of energy means just that: if you pump energy into something, it will retain that energy wherever it goes. To evaporate water, you heat it, either by placing it out in the sun or by applying heat directly from a stove. The more heat, the more evaporation. But what happens to the heat? It's gone, but not forgotten. In fact, it's retained in the water vapor as latent heat of vaporization, and will be released again when the water vapor undergoes the reverse process—condensation. So the tropical ocean and atmosphere work together as a huge heat pump, gathering the sun's energy over vast seas and pumping them into the center of the storm that is beginning to take shape.

Similarly, air conditioners work as heat pumps, moving heat from one place to another. To heat a room, heat can be pumped in from another location or it can be heated directly by building a fire. But cooling a room cannot be accomplished directly: there is no cold equivalent of a hot fire. In fact, there is no such thing as cold, there is only the absence of heat. So, to cool off a room, the heat must be pumped away. Air conditioners do this by evaporating a substance, which absorbs heat from the surroundings (the substance's latent heat of vaporization). The resulting vapor is then pumped outside and the substance condensed, releasing the absorbed heat. The condensed liquid is then cycled back inside, where it is again vaporized, and this continual cycling produces a cooling effect inside and a heating effect outside.

The big trick in the invention of air-conditioning was to find the right substance, one that would have good properties of evaporation and condensation at the temperatures needed, and would not have any serious deleterious effects. At the beginning of this century, when the work was first started, methyl chloride was used. It had good air-conditioning properties, but it was slightly corrosive and very flammable; the result was that it sometimes leaked out of the pipes, and if there was a fire in the house—such as a pilot light on a stove, or a fire in the fireplace—the house could explode. In the 1920s and 1930s there were more than occasional reports of such accidents from the new electric refrigerators, and so the world was thrilled in 1928 when Thomas Midgley discovered that a particular class of compounds had excellent refrigerant properties and were perfectly safe, totally harmless: the chlorofluorocarbons.

It took another fifty years or so before we discovered that these things eat up our protective ozone layer. But so it goes.

7

Meanwhile, back at the ranch, rising air masses are cooling off and losing their ability to hold water vapor. The molecules come together and condense into droplets. The rising mass hits the tropopause and begins to spill over, forming huge anvil-head clouds full of rain. These clouds are also full of energy, because the water vapor's latent heat of vaporization—originally supplied by the sun to the tropical oceans and since held by the water vapor—is released again when the vapor undergoes the reverse reaction of condensation.

What is happening is that the ocean waters have evaporated into the converging winds because of all the heat energy the sun poured into them; when they evaporated, they brought the energy along with them. Now as the air rises and cools and the water spills out into liquid droplets, this latent heat is released and the developing thunderstorm uses this tremendous release of energy to power itself into a lightning-throwing monster.

Thunderstorms develop all along an easterly wave, giving rise to a system of disorganized but powerful storms; the whole mess is called a tropical depression. All the processes so far described are normal and ordinary; so ordinary that no one bothers to pay much attention to them. Tropical depressions are not named and not normally reported in the newspapers or on television weather reports; they are too common to be news. In almost every tropical depression nothing further happens, as the thunderstorms blow themselves out. The water they picked up is rained back into the oceans, and nothing much has happened.

But once in a while a coincidence or two leads to this system of thunderstorms becoming something more.

8

We talked earlier about what happens when a mass of air begins to contract, and how conservation of angular momentum makes it spin counterclockwise. If for some reason the air packet expands, the opposite will happen: it will spin clockwise, becoming what is called an anticyclone. This can happen if a high-pressure area builds up: the high pressure at the center forces the air to expand and spread outward. This is one of the things that just sometimes happens, for no apparent reason. For a long time many scientists knew about it but paid little attention, because an anticyclone does not have the potential to develop into anything much at all, let alone a hurricane. Hurricanes develop because the conservation of angular momentum necessitates a speeding up of the winds as the mass contracts; it is this acceleration that causes the winds to reach hurricane intensity. An anticyclone, on the other hand, spreads out: its wind speed must decrease as its distance from the center increases.

This doesn't appear to be a big deal and, indeed, it wasn't until 1985, when a graduate student at Colorado State University, Robert Merrill, showed that if an anticyclone formed at high altitude, and by coincidence happened to sit directly above the center of a low-pressure tropical storm, all hell would break loose. When this happens a synergistic system is created: the upper-level high-pressure area in the center pushes the air away, and the low-pressure area at sea level sucks in air and sends it skyward into the center of the anticyclone, which then continues to build pressure and spin the air away. Consequently, the anticyclone sucks up more air from underneath and throws it well beyond the borders of the storm below.

Again, it is not terribly unusual for either of these things—the upper-level high or the lower-level low—to occur in subtropical regions. It *is* unusual for the high-altitude anticyclone to sit directly above the low-altitude cyclone. When this happens, the tropical depression has the potential to organize itself.

In a typical thunderstorm rising air bumps against the tropopause and spills over into a familiar anvil shape. It has now lost its water and cooled off, so it descends again to the surface, is sucked back into the ascending surface air, and is recycled. But the air being sucked back up has originated near the center; it has not traveled horizontally over the warm ocean surface but has descended vertically. In short, it has had no opportunity to recharge with more water vapor. And so when it rises it has no latent heat to contribute to the energy system, and as this process continues the thunderstorm runs out of energy—runs out of fuel, in essence—and dies out.

This is the normal course of events. But if a high-altitude anticyclone draws in the air ascending in the center of the storm and spreads it out with its spinning motion, a decidedly different storm is formed. Now the cool high-altitude air, which has already lost its evaporated water and thus donated all its energy to the developing storm, doesn't immediately fall back down to be recycled but is instead flung out hundreds of miles away. This lowers the pressure in the center of the thunderstorm and allows more warm surface air, saturated with water vapor, to rush upward. This new batch of air in turn contributes more energy as it dispenses its water vapor, and the original ascending air descends far away and is then drawn back in toward the center, traveling over the warm oceans and accumulating additional moisture. When it reaches the center and begins its new ascent, it contributes a new burst of fuel, and the energy of the storm builds up, and up, and up. . . .

With this sort of chain reaction, the system is now called a tropical storm. Out of the many tropical depressions, ten or more tropical storms may develop each year in both the Atlantic and Pacific.

What happens next is in the lap of the gods. We've seen by now that the air masses are pushed around by prevailing winds, and the winds can vary with altitude. In a Hadley cell, for example, over a typical point on the surface of the earth, we have winds blowing north at high altitude and blowing south at the surface. Every day we see examples of this divergence. Lying on the beach you might feel a breeze coming in off the water, while up above the clouds are scudding out to sea. This is because the winds are governed by different processes. Down at the level of the beach, the sun is heating the sand, which in turn emits infrared radiation and heats the air above it, which then rises. Over the ocean this process is not as effective: the ocean surface can reflect a lot of light, and some of the energy that is absorbed is transported down instead of being emitted back up into the air. So the air above the waters remains cooler and heavier than that over the beach, and when the warm beach air rises, the cool ocean air rushes in to take its place. Up at the level of the clouds this mechanism has died out, and something else is pushing the air around up there, perhaps in the same direction, perhaps not.

We see this sort of behavior everywhere. At ball games the wind can be fluttering the flags up high while the game is being played a hundred feet lower down in motionless air. A high pop fly sometimes seems to take on a life of its own, with the catcher staggering about wildly below trying to follow it as it encounters different winds at different heights. A northeasterly storm can be blowing in over Boston, while the jet stream higher up continues to sail from the west.

And so what we have is a crapshoot. A low-pressure cyclonic storm system is building over an easterly wave, and while the high-pressure anticyclone sits right over it and sucks up its ascending air and throws it out far into the Atlantic, the storm will continue to intensify. If a new, high-altitude wind blows the anticyclone in a different direction than the cyclone below it is being blown by the prevailing surface winds, the storm will subside. Only if the high- and low-altitude winds happen to stay in synchronization—and there is no mechanism for their doing this—will the storm continue to intensify. Once in a while this will indeed happen, just by chance. Then the cyclonic winds build up as the heat engine gathers the sun's energy from a vast oceanic area and pumps it into the center of the storm, and the high-altitude anticyclone spins the depleted air far out over the warm tropical oceans to be recycled. And when the converging wind speeds reach seventy-five miles per hour, the people at the National Hurricane Center in Florida pick up the telephone and say, "We now have a hurricane."

3

DIRECTION,
WHAT DIRECTION?

Maybe if you only knew
Where your path is leading to,
You'd become less wild . . .

—from "Dance, Little Lady,"
by Noel Coward

1

From *The Miami Herald,* Saturday, August 22, 1992:

> Tropical Storm Andrew, a distant swirl of wind until Friday, is scheduled for promotion this morning. Its new title will be Hurricane Andrew.
>
> "We don't like it either, but it's there and it looks like it's getting stronger," said Max Mayfield, a National Hurricane Center specialist who forecast Andrew's growth Friday.
>
> Something else to dislike: the latest forecasts have the storm changing its northwest direction to westward—toward South Florida, but slowly, and [it is] not likely to arrive before Tuesday.
>
> That doesn't mean it will happen. . . . "A lot can happen between now and next week," Mayfield said.

2

By Friday night the late-news announcer was saying that Andrew was scheduled to be upgraded to hurricane status by Saturday and was going to hit us within a few days. They cut away to clips of local food stores, with shots of people standing in long lines buying batteries, canned food, jugs of water, masking tape, and candles.

"Maybe you better go out and get some supplies," my wife suggested.

"I don't think so," I said. "It's not going to hit here."

"The man says it is." She pointed to the set.

"He doesn't know anything, he's just trying to drum up some excitement."

"You're not just saying that because of the last time?" my wife asked.

I wasn't, I assured her. The "last time" was nearly ten years before, when Hurricane David came roaring right up to us. It was supposed to hit at six in the morning; the day before, I loaded us up with supplies, and spent the evening tightening up the house and battening down the hatches while my children followed behind every step, laughing and getting in the way and taking pictures as I plastered masking tape over the windows to keep them from shattering and propped wood up against them to keep them from blowing in, and brought in all the outside furniture and filled the tub with water. I climbed up on the roof and took down the TV antenna and unscrewed the rotary fans and brought them in. When I finished, the entire inside of the house was filled with all our outside furniture; in Florida we live mostly outdoors, and have more stuff outside than in, so when I brought everything in it was like living in a secondhand furniture store preparing a going-out-of-business sale. Then I pulled the mattresses off all the beds and brought them into the inner hallway, so if the windows shattered despite my precautions, my children would still be safe from flying glass. And all the while they laughed and giggled and took pictures.

When I finally got them settled down for the night the wind was beginning to blow in earnest, and I sat up all night listening to it build, wondering if the roof would hold. Finally I fell asleep; the last time I looked at the clock it was four o'clock.

I woke at seven-thirty, and the house was quiet. There was no sound of wind. I looked out the windows; it was a cloudy day, but that was all. I turned on the TV, and found that at the last minute David had swerved sharply to the north, and was heading back to the open Atlantic.

My kids woke early enough to take more pictures of me taking the outside furniture back outside, putting back the TV antenna, reconstructing the rotary fans on the roof. . . . They still love to show those pictures.

"It has nothing to do with last time," I told my wife. After all, my kids are all grown and on their own by now. But I had learned a lot about hurricanes in the past ten years, and had a better understanding of how and where they moved than the pretty boys on television had. "This one won't hit," I told her. "Not here, anyhow. Hurricanes always curve up to the north, and this one is just about at our level; when it curves north it'll miss us."

At that moment Andrew was at latitude 25.6 degrees and longitude 66.5 degrees, 852 miles due west of Miami.

3

Saturday morning I got up early and picked up the paper first thing, and found Andrew was still headed west, right at us. "Damn," I swore to myself. I just knew it wasn't going to hit us, but I also knew, deep down, that I didn't know anything for sure.

The larger problem was my steady Saturday-morning tennis game that I didn't want to miss, and, on top of that, I was near the end of a series of mass-spectrometer measurements that would tell a pretty story about how the atmosphere evolved. I needed one more good experimental run to confirm what I thought I had discovered, and I was set up to do that the next workday, ordinarily Monday morning. The hurricane was scheduled for Tuesday morning.

It probably would not hit, almost certainly not before Tuesday, but if it did it would knock out the electricity at the lab. The mass spectrometer and its pumps are kept working continuously; if the power is interrupted, there are all sorts of subtle shades of meaning in its innards that might not come back to exactly the right nuance, and I'd have to start all over again, a loss of several months work. Andrew probably would not hit, I told myself again, and even if it did I would probably have time to do the run Monday . . .

The more I thought about it, the less good that word *probably* looked. I called up my tennis partner and cancelled, ate breakfast, and drove out to the lab. I spent the day there completing the run, and it turned out perfectly. I came home with all the satisfaction of having done something right; it's not a feeling I'm used to, so I really enjoyed it.

Sunday morning I woke up early and turned on the radio. Andrew had sped up during the night while still maintaining its westward course. It was supposed to hit us Monday morning instead of Tuesday.

I didn't believe the radio. Who listens to disembodied voices? I got out of bed, went into the family room, and turned on the television set, hoping to find somebody who might mention the storm. Not to worry: every station was talking about nothing else. It was heading directly for us, it had not swerved the least bit for three days now, it was bearing down directly on us and going like hellfire.

I didn't believe it. Hurricanes curve northward, or at least 99 percent of them do. The probability that this one would not, and that it just happened to be the one heading for *me*, was too coincidental to be believed. I opened the door and looked outside. Although the television was insisting that Andrew was coming right at us, the weather was absolutely perfect. I didn't want to miss my Sunday morning basketball game, but the television people kept harping on about Andrew no matter how many stations I turned to. Reluctantly, I put on sandals instead of my Michael Jordan high-tops and went out to pick up some supplies at the supermarket, just in case.

Pulling into the parking lot I was astonished: it should have been empty this early on a Sunday morning, but it was mobbed. It was bedlam. There were no carts available; a line of people waited to follow shoppers out to their cars to claim empty carts. By the time I got one and got back inside there was nothing left on the shelves. I picked up a can of tomato soup—useless, for we have an electric stove and electricity would be the first thing to go—and, luckily, some toilet paper, the last two packages. I got back into the car and cruised past a couple of 7-Elevens, but it was obvious from the crowds around them that I wouldn't do any better there. I went home and ate and told my wife what was happening. "It's just panic," I assured her. "It won't hit us. Probably," I added. She didn't like the sound of that word any better than I did. After breakfast I drove out to the laboratory to tie down my equipment, just to be on the safe side.

Our lab, the Rosenstiel School of Marine and Atmospheric Sciences, is located on Virginia Key, an island off the coast of Miami that is connected to the mainland by an artificial causeway named after Eddie Rickenbacker, whom most people think was just president of Eastern Airlines, of blessed memory. Sitting out there in the middle of the Atlantic waters, we are prime candidates for total destruction in a hurricane. At about a foot above sea level, any decent storm surge will wash right over the whole island.

Well, nothing to be done about that; you can't pack up and move all the equipment. So I did the next best thing: I began to seal the cracks around the doors and windows with Scotch tape. Everyone else was there, too, sealing up against windblown water as best as they could, but laughing at the uselessness of it all. "There's no way this is going to hit us," we all agreed. Because we knew how these storms are guided by the prevailing air currents, we understood about high- and low-pressure areas and their effects; we knew the historical record. There were anomalies in the record, to be sure, like the 1965 storm named Betsy that wandered around as if she were doing a waltz: left, together, right, together, and then one step backward. But the paths of such storms were just that: anomalies, not to be repeated. "No way," we all agreed. Taping up doors and throwing plastic over all the equipment was just good practice for the hurricane that was sure to hit some day. Some day, but not this day.

Because we knew how hurricanes move.

4

It was Benjamin Franklin who first reached the seemingly simple but previously misunderstood conclusion that storms do not just happen, that they do not spring full-blown from the forehead of a primeval god or step out of an opened oyster shell; instead, they grow slowly into maturity while moving from their origin to wherever it is they're going to cause trouble. The reason this was hard to understand is not hard to understand. When a hurricane suddenly came blowing over the horizon in those bygone days, it came without warning as a suddenly pouncing monster. It came scudding over the waves with black clouds and boisterous winds, exploding on the unsuspecting like a gigantic bomb. It seemed reasonable to think of the storm as something that just suddenly happened, like other natural catastrophes of those days: pregnancy, the pox, drought, and famine.

This view persisted until one night when Ben Franklin went outside to observe an eclipse of the moon, and was disappointed. A storm obscured the sky, and although he stood in the rain until he was soaking wet and had to go indoors or die of ague, the moon was not to be seen. (Or rather, the moon that was not to be seen was not to be seen.) Initially he was disappointed but later invigorated when he read in the Boston newspapers that the eclipse had been visible there that night, although the following evening a terrible storm had hit. It occurred to Franklin that the Boston storm was actually the same storm as the one in Philadelphia, and that it had *traveled* from Philadelphia to Boston.

This seems so obvious, it's hard to realize that no one before him had thought of the possibility of storms traveling from one spot to another instead of exploding spontaneously in the region where they were observed. But genius, my grandfather used to say, was simply the knack of looking at something everyone had looked at before and seeing something no one had ever seen before. The idea so excited Franklin that he sat right down and wrote letters to people up and down the thirteen colonies, and found out that a similar storm had hit Virginia the day before, South Carolina before that, and New York in between Philadelphia and Boston. With that news the science of meteorology was born.

He conceived the idea of weather maps, of plotting all the storms that existed on a given day and following them on succeeding days. Given such a set of data, it would be easy to predict where each storm would be the following day. For instance, if you knew where a storm had been on Monday, Tuesday, and Wednesday, you could say with some certainty where it *had* to go on Thursday.

The great difficulty, Franklin soon discovered, lay in the innocuous phrase *given such a set of data.* How to gather such data? In fact, it turned out to be impossible. If a storm hit Atlanta, his correspondent there would immediately notify him—by post. The man would write a letter and it would be put on the stagecoach, and perhaps a week later it would be received in Philadelphia. The problem was that the storms moved faster than the postal system. (Today it's hard to be surprised at that, for now *everything* moves faster than the postal system.) To chart storms at sea was even more difficult. A ship might be in position to observe the storm, and the captain might be disposed to pass on the information, but there was no way he could do it until they docked, perhaps several weeks later. And so it was impossible to gather together the necessary data set within the time constraints imposed by the movements of the weather systems.

It remained impossible for more than fifty years, until 1832, when Samuel F. Morse was returning to America from Europe on the packet ship *Sully.* Several of the better-educated gents aboard were discussing some recent work of the English scientist Michael Faraday, who had demonstrated how to create an electric charge by moving a magnet back and forth within the coils of a metal wire. A Dr. Watson of Boston remarked that if one could create an electric charge, it ought to be possible to pass it along the wire from one place to another. Morse then thought that if one could also *detect* such an electric charge, one could send signals by opening the circuit, to stop the current, and closing it to allow the current to flow again. By the time the *Sully* arrived in New York he had the details worked out and had invented the dot-dash system for converting the resulting signals into letters.

It took another ten years before he was able to convince Congress to appropriate thirty thousand dollars to construct an experimental telegraph line stretching from Baltimore to Washington. The experiment was a success, the line worked perfectly, but when Morse offered his system to the government for $100,000, it was turned down on the advice of the postmaster-general, who was unsure about the demand for telegraphic services and "uncertain that the revenues could be made equal to its expenditures." Morse then raised private capital, and by 1851 fifty companies were operating his system throughout the United States. Three years later there were several weather stations operating in the eastern United States that reported daily to a data-collection center at the Smithsonian in Washington, and meteorology had passed through its infant stage into puberty. (In Europe the baby was a bit older, and a bit premature: a series of observing stations had been set up in France, chiefly by the Chevalier de Lamarck—from 1800 to 1815—prior to Morse's invention. In 1820 a professor at the University of Breslau, H. W. Brandes, produced the world's first daily weather map—but it was for the year 1783. The establishment of current weather maps had to wait for the introduction of telegraphy.)

The transition into puberty was a difficult one; well, it always is, isn't it? In 1834 Pennsylvania appropriated four thousand dollars to fund one weatherman with a barometer and a thermometer. By 1842 the federal government was hiring a meteorologist or two. One of the first was James Espy, who made it his life's mission to begin a national weather service. It was not an easy battle: "Mr Espy . . . is methodically monomaniacal," John Quincy Adams wrote. "The dimensions of his organ of self-esteem have swollen to the size of a goiter by a report . . . endorsing all his crack-brained discoveries in meteorology."

It was Espy who lobbied for the establishment of weather observatories throughout the country, with a national data center in Washington to which they would all report and which would analyze and interpret the received data. It was his contention that he could make sense of all these disparate data—that the weather in each individual city was related to and formed by a continental weather mass—that constituted his major "crack-brained discovery."

But despite Mr. Adams's opposition, Espy's federal weather service eventually came into being, and the science of meteorology was alive and kicking. It came into maturity with the spread of weather stations across the globe, with direct radio communication to ships at sea and planes in the skies, and finally with weather satellites, which monitor the weather all over the world at every instant.

So why do we still not know what's going on? Why is it that you listen to the television weather report at night and are told the skies will be sunny tomorrow, and you step outside in the morning and get drenched with rain? It's not simply occasionally that weather reports are wrong; it's pretty much the usual thing. As a result, many stations hedge their bets by predicting something like "The chances of rain tomorrow are thirty percent." What does that mean? Does it mean that it will rain over 30 percent of the viewing audience, storming perhaps in north Miami while the sun shines in south Miami? Or does it mean that there's a 30 percent chance of it raining *somewhere* in the area? Or that there's a 30 percent chance of rain covering the entire area and a 70 percent chance of clear skies throughout?

It means that the weatherman can't be wrong, and that's all it means. If it rains anywhere, well, fine, he told you so. If it doesn't rain anywhere, well, he told you that, too. It means that we can't predict the weather, that's what it means.

This has serious consequences, much more serious than going to work without an umbrella and getting soaked, for the world today is on the verge of an atmospheric disaster that could be the worst thing in our collective history. Yet most people don't believe it, suggesting that if we can't predict tomorrow's weather, how can we predict what it will be like in fifty years?

The catastrophe is known as global warming; it is caused by the increase in carbon dioxide from burning fossil fuels, together with a host of secondary associated problems such as the cutting down of the world's forests and a concomitant increase in methane and fluorocarbons. Because these gases are all greenhouse gases, the earth will get warmer in the future if we don't stop increasing their concentration in the atmosphere.

This much is absolutely known, is absolutely true; but much else is not. It is not known how great the rise in temperature will be with any given increase in CO_2 or methane or chlorofluorocarbons, nor how fast the rise in temperature will occur, and it is not known how deleterious—or perhaps how disastrous—the associated effects, like a rise in sea level, will be. All these uncertainties remind people how uncertain daily weather reports are, and they therefore ask —reasonably enough—how can we be so sure of global warming at all? If we can't tell you whether it will be hotter tomorrow, how can we tell you it will be hotter fifty years from now?

Well, I may not be able to tell you tomorrow's weather, but I can tell you next year's climate. For instance, it will be hotter in New York next July than it will be next February, and it will be colder in Maine in January than it will be in Morocco. I can tell you that Florida will get more rain next year than the Sahara will.

Large-scale weather patterns that govern climate can be predicted with confidence, even though the smaller-scale patterns that determine daily weather cannot. The reason for this is simple: it has nothing to do with our basic understanding, but with the number of complicating factors. The bigger a weather pattern is, the more stable it is, and the less susceptible to small disturbances. Yearly and global climates are very steady, and the effect of a single large disturbance on them, such as the increase of greenhouse gases, can be predicted. (Even in this case, however, we can't predict with accuracy just how fast and how large the temperature increase will be.) To move from considerations of climate to discussion of the daily weather means going from long-term stability to short-term chaos. Last night, for example, as I wrote the preceding pages in a small cottage in Maine on January 14, 1993, a blizzard was developing offshore. The prevailing winds were northeasterly, meaning the storm was blowing right into us. The weather report predicted up to two feet of snow by morning, and continuing throughout the day. So we went out and stocked up on enough food to get us through a few days of isolation, and worried all night how we could heat the place if the electricity went out. In the morning we looked out on clear blue skies, the most beautiful day in months—resulting from a high-pressure area to the northwest that overpowered the easterly winds and blew the blizzard back out to sea. This could not have been predicted, because the high-pressure area should have drifted too far south to affect the storm, but a shift in the jet stream over Canada brought it back, the jet stream's shift being caused by another disturbance over the Pacific, which in turn was caused by—the beating of a butterfly's wings in Vietnam?

The point is that the smaller the weather system, the more things there are that can affect it. Daily weather is affected by so many interactions that even with the help of computers, meteorologists are quite helpless to predict what is going to happen. Yet with larger weather pictures, such as global climate, the disturbing factors are small enough in number to be handled, although more qualitatively than we would wish.

Hurricanes are between these two extremes. The beginning of a hurricane is so delicate that it can be magnified or canceled by any of a dozen causes, making any prediction a wild guess. But once a monster storm gathers itself together we can predict—at least to some extent—where it will go next, because not only is it a large-scale system but it is directed by an even larger and more permanent condition.

5

A sailing ship on the high seas will move along as the wind pushes it, except that sailors have learned how to tack against the wind, using the pressure of the keel against the slower moving water to provide a fulcrum against which to push. A mass of air, such as a hurricane, has no keel and will therefore move wherever the steering currents push it.

The steering currents consist of the large air mass in which the hurricane is imbedded, together with the prevailing winds, which are dictated by other, even larger air masses. If the steering currents were marked in the air by arrows of different colors, as weather systems can be on the television reports, predicting the movement of the hurricane would be simple since, like a sailing ship without a keel or a balloon in the wind, it *has* to follow the currents.

But God didn't see fit to make things that simple, and defining the proper steering current for any individual hurricane is not always an easy thing to do, since air masses don't have definite boundaries and merge into each other as they swirl and wander about. On the other hand, there are a few large-scale and straightforward, rather permanent atmospheric features out there on which one can rely reasonably well.

To return to the Hadley cell (and its southern hemisphere twin), we notice that warm, moist tropical air rises over the equatorial regions and moves poleward. As it rises high it cools off and loses its moisture, which is why rain is such a typical tropical feature. By the time it's reached the northernmost limit of the Hadley cell, at about latitude 30 degrees, the air is totally dry. At this point it begins to descend as a converging mass of dried out air, warmed by the force of compression and a lower altitude. What is not present under these conditions is a lot of rain. This accounts for the great deserts of the world—the Sahara of North Africa and the Kalahari of South Africa; the Atacama of South America and its northern equivalent, the Sonora of Mexico; the Great Nafūd desert of Arabia and the Helmand of Asia—which are found beneath this air mass.

Furthermore, the continual accumulation of this converging air results in a series of high-pressure systems known collectively as subtropical ridges, which sit more or less in permanent place and are known individually by their geographic centers: the Bermuda high, the Azores high, and the North Pacific high.

When a hurricane develops to the south of one of these highs, the constant air pressure pushes it away. Combining this with the prevailing easterlies means a hurricane sails on the edge of the winds toward the west.

It crosses the ocean on the winds of the high, and by the time it reaches the West Indies it's running beyond the range of these high-born winds, so it curls around to the north along the edge of the high. Here it begins to encounter the mid-latitude westerlies, which turn it around and begin to push it in a northeasterly direction. The result is a roughly parabolic path, first to the west from its place of origin, and then up and around to the north, and back again to the northeast. The westernmost storm shown (see Figure 3-1) would swerve along the eastern coast of the United States and then move out to sea. If the high were a bit more extensive, or located a bit farther west than shown, the storm could slam into the East Coast before turning. If the high were even more extensive, or far-

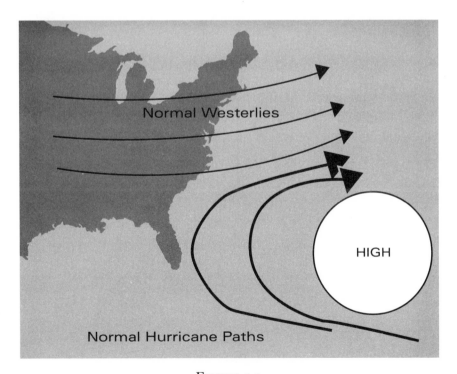

FIGURE 3-1

ther west, the storm would sail east past the southern tip of Florida before beginning its northward turn; it would then blow through the Gulf and hit land somewhere along the Gulf Coast.

Storms that miss America, like the ones illustrated, might slam into the coast of northern Europe on their return voyage save for the fact that the surface waters to the north get colder and, as less water is evaporated, the hurricane runs out of fuel. Occasionally one does make it all the way, such as the Great Gale of 1703, which hit England.

The Great Gale was an unusual storm in all respects, not only surviving from its origin eastward across the Atlantic and back again to the shores of Britain, but also occurring remarkably late in the season. On the afternoon of November 26, 1703, following several weeks of stormy weather, gale-force winds began to blow over the southwestern counties of England. There was, of course, no warning of a coming hurricane, and people battened down their windows and went to bed when night fell, fearing nothing more than just another storm. The night fell with uncommon darkness, for it was a night of the new moon, which in those years (before the advent of electrical or even gas lighting) meant a very dark night indeed.

The wind increased hourly, and soon bits of roofs were flying off. At St. James's Palace the Queen was in residence, and fled from her bed when half the roof was ripped off. The Bishop of Wells also got up at night when he heard the sound of structural damage; as he made for his bedroom door a chimney stack crashed through the roof, killing him. In all, eight hundred houses were demolished, and innumerable others were damaged. The church with the highest steeple in Kent was blown down: "This strong and noble structure by the rage of the winds was levelled with the ground, and made the sport and pastime of boys and girls, who to future ages can boast that they leap'd over a steeple."

On the southwestern coast the storm surge produced eight feet of water in Bristol, flooding all the cellars in that shipping town, destroying the rich stores of the West Indian trade. "They tell us the damage done by the tide amounts to above 200,000 pounds; 15,000 sheep drown'd in one level, multitudes of cattle on all the sides, and the covering of lands with salt water is a damage which cannot well be estimated."

As was to happen following Andrew, the price of roofing tiles tripled in England that year, and bricklayers gained a 150 percent raise in their rates. (We were to see much of that sort of thing in Miami in the next few months.) In all, some eight thousand lives were lost, many of them in ships at sea. So many, in fact, that the Royal Navy had to be re-peopled with convicts, thugs, and murderers, whose death sentences were commuted if they "volunteered" for Her Majesty's ships. "It's an ill wind that blows no one any good; and the Great Storm certainly saved a handful of rogues a hanging."

So we know how hurricanes move. But we didn't know how Andrew was going to move. Why not? For the same reason we can say we know that August and September are going to be wet months in Miami, but we don't know on which days it will rain. The general weather-circulation system is reasonably steady on a long-term basis, and the average motion of hurricanes will follow a given pattern. But the system itself can vary on a daily basis, and any given storm can move in ways mysterious to behold. Out at the marine school that Sunday, we forgot that.

6

On September 1, 1965, Hurricane Betsy was at nearly today's position for Andrew: just below latitude 25 degrees and at the same longitude, putting her just north of San Juan, Puerto Rico. She was following the classical hurricane track, heading due west but beginning to curve northward around the limits of the Bermuda high.

By September 5 she had progressed to nearly latitude 30 degrees north while moving to 75 degrees west, placing her about three hundred miles off the Florida coast, nearly even with Jacksonville.

She looked to be heading up to hit the coast, perhaps around the Chesapeake Bay or, perhaps, to miss it entirely and curve around and fade away into the North Atlantic waters.

But an unexpected low-pressure trough suddenly developed in the westerlies over the continental United States. This drew the Bermuda high to the west, placing it just north of Betsy, blocking her path.

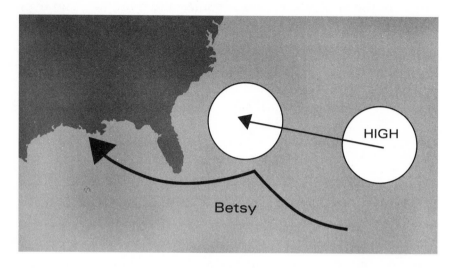

FIGURE 3-2

Betsy sat there for a day, not knowing what to do. The Bermuda high remained in its new position, so that the steering currents flowing from it bore down on Betsy from the north, causing her to turn around and blow south. The wind pattern reestablished itself around the new position of the high by the time Betsy moved down to 65 degrees, just below Miami, and so she turned westward again, sailed over the Keys and into the Gulf, and then, turning north again, moved around the flank of the high. A few days later she blew into New Orleans, hitting the city with only a few days' warning, since the week before she had been heading in a completely different direction.

Although Betsy is hardly unique in history, she does not represent the usual series of events. Ordinarily, hurricanes are quite well behaved storms: they move wherever the steering currents direct. And ordinarily, the steering currents' customary path circles around the Bermuda high, which ordinarily sits still and stolid in its customary place. Ordinarily, but not always.

So we should not have been so blasé on that Sunday afternoon as Andrew bore down on us. But he was still hundreds of miles away and due west of us. The smallest turn northward would divert him farther up the coast; a normal turn northward would curve him around and back out to sea. The odds of his continuing due west without any variation at all were unreasonably high. So I taped the door to my lab with Scotch tape, just in case Andrew came close enough to blow some rain at us, and I went home.

7

The only way to deal with the many and varied complexities that influence a hurricane's path is by computer modeling. You speak to the computer in its own particular language and you describe the world out there, and you ask the computer how hurricanes will move about in that world. But even with all the power of the world's fastest supercomputers, nature is still too complicated. So you have to simplify things a bit, and this is where problems arise. Because every simplification is, of necessity, a departure from reality.

There are about a half-dozen groups of people across the country doing this kind of work, and while they agree on many things, they disagree on a lot of others. In particular they disagree as to which simplifications should be made. One group thinks that process A is likely to be unimportant and can be represented by a constant, while another group thinks that A might be the key to the whole thing and must be looked at in detail, with every possible variation; this group thinks that process B is what can be simplified, while another group thinks that B is the key to everything, and C might be simplified. . . .

These groups get together and talk at scientific meetings, visit each other's laboratories, and try to convince each other of their arguments. At present, no one has convinced any of the others that their particular model is superior. When Andrew had finally passed over, when the worst of the damage had been repaired and we were beginning to put our lives together again, the entire faculty at the Rosenstiel School had a meeting on Virginia Key. Dr. Rainer Bleck, the head of the division of meteorology and physical ocean-ography, took the floor and tried to explain what had happened. Someone pointed out that on Friday morning the predicted track had indicated Andrew would still be east of the Bahamas on Mon-day afternoon, whereas it hit Miami Sunday night. The first real warning to Miami came only thirty-six hours before it hit, and even then it was not much more than a vague hint. Why weren't we given better warning of what was happening?

The answer is rather easy, Dr. Bleck said. No one could have given better warning, because no one knew what was happening. He explained that in retrospect he did have a computer model that predicted Andrew's path perfectly, but at the time he couldn't know how well the model would work. Further discussions brought out the fact that in addition to this model which gave the right answer, there were four others that gave different possible paths, and re-sults from previous hurricanes had acknowledged none of the five models as superior to any of the rest. Even today there was still no reason to think that the model which worked for Andrew would do any better than the others for the next hurricane. It's like talking to five different touts at a racetrack, each one giving an inside tip on a different winner. When one of them wins, you kick yourself for not having listened to that particular tout, but if you listen to him next time, you'll probably lose your money.

The difficulty lies in the vast number and large magnitude of complicating and indecipherable factors. If a hurricane is pictured as a cork bobbing in the current, it seems easy enough to predict where the currents will take it. But it's more like a sailing ship, for a cork responds only to the movement of the water, while a sailboat responds to the wind in its sails as well.

And it's worse than that. The hurricane is in some ways like a cork, in others like a sailboat, and in others more like a swirling eddy in the current itself: its own motions feed back into the current and affect it, changing its direction or its strength. And the steering current itself is more like a living, growing, writhing, fornicating sea monster than like an inanimate, simple wind. The difficulty of making predictions can be hinted at by taking a closer look at this wind.

Instead of blowing steadily along, the steering current is subject to the same forces that shape the hurricane. In particular, it can begin to spin itself, if, for example, a low-pressure area forms to one side of it or a high-pressure area to the other. Should it start to spin clockwise to one side of the embedded hurricane, the hurricane will move to the other; if it spins counterclockwise the storm will move to the same side. Further, if there is a convergence upstream or a divergence downstream of the steering current, the hurricane will speed up, as if a river were flowing down a mountainside or into a waterfall, while if there is a convergence downstream or a divergence upstream, it would have the opposite effect.

It's not only external factors that affect the steering current, the hurricane itself is massive and energetic enough to do so. For example, we've seen how the ascending air in the storm's center is spun out to great distances before it sinks again. Surprisingly, this outflow is not circularly symmetric, but tends to concentrate in certain directions, which creates jets of spurting air. The factors influencing the precise location of these jets are not yet understood, and thus can't be predicted but only monitored. They change with time, as the hurricane grows and forms, and like rockets on a flywheel they can change the spin and orientation of the hurricane within the steering current, and ultimately the direction of the storm. If strong enough, they can even change the direction of the steering current. This factor is just beginning to be appreciated; as recently as 1988, it was ignored by forecasters predicting the path of Hurricane Gilbert (their warning to the Texas coast did not exactly prepare the region of Mexico that got hit).

. . .

On Sunday, the National Hurricane Center in Miami finally decided to announce that Andrew was going to hit us, and that the coastal areas should be evacuated. My father lives in a high-rise condominium in north Miami, a few hundred yards from the beach. He called to ask what he should do, and I told him to drive down to us right away. He said that he was uncertain: most of his friends were staying put.

The condo he lives in isn't exactly a senior citizens' home, but it's pretty nearly one. The residents are mostly retired New Yorkers or Philadelphians, and they can be pretty stubborn. I told him not to fool around: he had to leave, and he ought to convince his friends to go, too. His building— *all* the buildings along the beach in that area —were built within the last two or three decades. Dade County, in which Miami is embedded, has a construction code practically guaranteed to make buildings hurricane-proof, but we hadn't had a hurricane since 1966, and that was a small one. There had hardly been a big one in living memory, and I hate to be cynical or mean-spirited but I just could not see all those Florida building contractors obeying the construction code every year and spending money to make buildings safe against a hurricane that hadn't come since before God-knew-when and wasn't going to come again till even God-didn't-know-when. . . .

I had difficulty trusting the contractors and building inspectors, and I didn't think any of those buildings would remain standing if the hurricane actually hit them. I told my father to come south, to us, as quickly as he could.

4

IMPACT

You do something to me
That nobody else could do . . .

—from "You Do Something to Me,"
by Cole Porter

1

From *The Miami Herald,* Sunday, August 23, 1992:

BIGGER, STRONGER, CLOSER
South Florida Bracing
for Hurricane Andrew
The realization hits home:
This one really could be it

Hurricane Andrew grew in strength and speed Saturday night. Forecasters issued a hurricane watch from Key West to Titusville, as the storm chewed its way across the Atlantic toward the heavily developed South Florida coast.

Andrew had sustained winds of up to 110 miles per hour—making it almost a Category 3 hurricane, which can bring along a storm surge of 7 to 9 feet above high tide.

Civil defense coordinators gathered in thick-walled bunkers and issued the following advice: "Above all, do not panic . . ."

2

Hurricanes are basically gigantic heat pumps, gathering in the sun's energy over a region of hundreds of square miles and pumping it into the small center of the storm. The mechanism they use is evaporation of the warm tropical seawater, which transports the energy, and its subsequent condensation, which releases the heat into the interior. Like any engine, for it to keep running there has also to be a way to get rid of the exhaust: in this case, the dry air that results from the loss of water vapor. The mechanism for this is a high-altitude anticyclone, which sucks up the central dry air and spins it out hundreds of miles away. Air pressure is lowered in the center, where the warm, wet air is being sucked up to be spun away at high altitude, and with a low-pressure center, more air rushes in to fill the partial vacuum. Because of the earth's rotation and the storm center's position above (or below) the equator, the inrushing winds are spinning around the center as they try to move into it. The greater the difference in air pressure between the normal, outside air and the partial vacuum inside, the greater the velocity of these winds.

As the storm moves across the waters it can grow or weaken, depending largely on the coordination between inrushing surface winds and outpouring high-altitude winds. Since each of these is governed by separate steering currents—with wind direction and velocity changing drastically with altitude—it's very difficult to predict just what is going to happen from day to day. The hurricane can straggle along and never become much worse than a bad thunderstorm, or it can grow until it roars into a city with absolutely devastating force, killing more people and causing more damage than anything else in all of nature's arsenal. It can be more deadly than any earthquake or volcano, more devastating than any forest fire or mudslide. The hurricane that blasted Bangladesh in 1970 killed so many people that only rough estimates are possible, and these range up to 1 million dead, while Andrew was to cause more damage in terms of material goods than any natural disaster in world history.

3

There is a world of difference between one hurricane and another. The storm that hit Miami in 1966 did little more than rattle a few windows and dump a few inches of rain on the Everglades. The hurricane that hit Galveston, Texas, in 1900 killed nine thousand people.

To bring order out of this chaos Admiral Sir Francis Beaufort, hydrographer of the Royal Navy, proposed in 1806 a classification system for storms of all types. It was a simple numerical system running from zero to twelve, based on the force of the wind. Calm air, in which smoke rises vertically, was ranked as 0. When smoke began to drift, it was a 1; if you could feel the wind in your face, it became a 2. When leaves began to blow about, you had a force 3 wind; when dust and papers began to move around, you were in a force 4. Small trees sway in a force 5 breeze, and by the time it gets hard to keep an umbrella from blowing inside out, the wind is up to a 6. It becomes hard to walk against a force 7 wind, and an 8 begins to break branches off the trees. Bits of roof come off in a 9, trees are uprooted by a 10, widespread damage accompanies an 11, and with a force 12 you have a hurricane: "More or less complete destruction."

The original scale, as written above, was given in terms of typical observations a person might make rather than in terms of actual wind speed, because there was no way to measure wind speed in 1806. Moreover, estimates of the strength of the wind became more difficult to make in following years as steamships replaced sailing vessels; in the latter you are moving more or less with the wind and can feel its effects, while in the former sailors began to move through the wind without respect to its force, and had to estimate wind speeds by observing waves and spray. Not until 1846 was an accurate method of measuring the speed of the wind invented, when J.T.R. Robinson, an Irish astronomer, presented his anemometer. It was a simple instrument, consisting of four hemispherical cups mounted on arms connected to a central rod. The wind spun the cups, which spun the arms, which spun a dial, which measured how fast the cups were rotating. With this advance it became possible to quantize the Beaufort scale, and we now call a force 12 wind anything over seventy-five miles per hour; that number becomes the defining criterion for a hurricane.

Wind speed is the most obvious thing one can measure in a hurricane, and accordingly it was the first. But what you really want to know about a hurricane is a bit more than that. Scientists studying the storm want to know about its structure and development, and people in its path want to know how severe it will be. We can get a handle on both these questions if we know how far the pressure at the center is dropping, because it is the difference in air pressure between the center and the surrounding areas that drives the winds. Air pressure, in fact, is the determining factor in all weather prediction. It's a concept that was recognized only relatively recently, although the instrument that measures it was invented in 1643.

In that year an Italian named Evangelista Torricelli conceived the idea that an empty room is not empty. Although the true nature of air was not demonstrated until 1783 (by Lavoisier), 141 years earlier Torricelli was convinced that the nothing all around us must consist of something. His reasoning is not known, but we can make a guess at the kind of observations that stimulated him. Consider a flame, for example. Put a glass over a lit candle and the flame will soon die out, suggesting that there must be something in the air that feeds it. Since this is demonstrably so, the "air" must be something, cannot be nothing.

Interesting idea. Others must have had similar thoughts, but what set Torricelli apart was that he devised a way to prove it. His idea was simplicity itself. Consider a hollow tube bent into a U shape and filled with liquid, on the left in Figure 4-1.

The liquid will obviously be at the same level in a and b. Even though this was intuitively obvious to everyone before him, it was Torricelli who suggested that it was due to the identical weight of the air above the liquid in the two places. Sealing off the tube on the

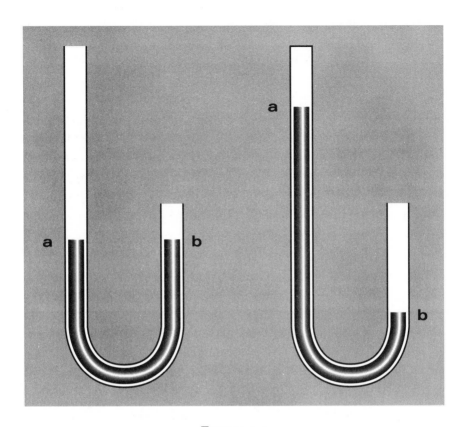

FIGURE 4-1

left side *(a)* and pumping away the air above the liquid will leave no air to push down on *a*, and so the weight of air above *b* will force the liquid down.

And that is exactly what happened when Torricelli constructed his "barometer." Under normal atmospheric conditions the level of mercury at *a* will be seventy-six centimeters higher than at *b*. (Mercury is used because it is a liquid metal and thus has a low vapor pressure; that is, few atoms break away from its surface and fly around in a gaseous state. If a normal liquid such as water were used, the space above *a* would fill with a significant vapor pressure, which would then push down on the *a* column and disturb the reading.) If we imagine a column of air above the open inlet at *b*, there is a certain number of molecules in that column, each weighing a small but finite amount. At sea level the air in a column above a one-square-inch opening weighs 14.7 pounds, enough to push the liquid down seventy-six centimeters below the level at *a*, where there is no air.

In Torricelli's day this was not so obvious, since no one believed in molecules or atoms despite the fact that Democritus had hypothesized their existence more than two thousand years earlier. But by 1842 James Espy was arguing that the barometer was measuring the varying weight of the atmosphere, and it was this variation in weight (or pressure) that accounted for winds and storms. Presumably this was one of those "crack-brained discoveries" Mr. Adams so deprecated.

We know now that as air masses circulate over the earth building up areas of high and low pressure, molecules of air will stream from high to low pressure just as water will squirt from a nozzle because of the high pressure behind it and low pressure in front of it. This perpetual motion—fueled by the sun's uneven heating of the surface of the earth—provides not only our winds but all our weather. (The term "pressure" is used rather than "weight" because molecules of air are in constant motion, and as each molecule collides with something—which it does about 10 billion times a second—it exerts a pressure on it. Since this motion is random, pressure is exerted in all directions, unlike weight, which simply exerts pressure "down." It is this omnidirectional pressure that keeps us from feeling the weight of the atmosphere: the molecules inside our body are pushing out with the same force that the air molecules are pushing in; the air molecules under an extended hand are pushing up with as much force as those above it are pushing down.)

The winds of a hurricane are no exception to these laws, and are related to the difference in pressure between the center of the storm and the outside world by a simple equation:

$$Wind\ Velocity = 6.3\ \sqrt{Pressure\ Difference}$$

The pressure differential is related to the ocean surface temperature, and since the wind velocity is related to the wind energy and thus its potential for destruction, the damage a hurricane does is directly proportional to the pressure difference inside and outside of the eye. (It's intuitively obvious that the more energy a wind has, the more damage it can do; and you probably remember from high school that $E = \frac{1}{2}(mass) \times (velocity)^2$. Nor should the relation between ocean temperature and pressure difference come as a surprise: since the hurricane is driven by the solar-heat engine (the sun heating the ocean waters, which then evaporate and feed the storm), it should be intuitively obvious that the warmer the ocean temperatures the more energy is available for the hurricane, which means the faster the central air column will rise, creating even lower pressure there. ("Intuitively obvious" is an expression my physics professor in graduate school used to use, e.g., "It should of course be intuitively obvious that the total rate of radiation escaping from a black box is proportional to the fourth power of the absolute temperature." The expression drove me crazy then, but I find it wonderfully useful now.)

Robert Merrill, who in 1985 showed that a high-altitude anticyclone over the low-altitude cyclone was a necessary part of starting a hurricane, also provided an estimate of the maximum potential wind speed as a function of seawater temperature.

The graph (Figure 4-2) makes clear that in order to get a hurricane going, seawater temperatures greater than about 25 degrees Centigrade are needed, and the ferocity of a hurricane can rise steeply as the ocean waters get warmer degree by degree. This great variation in the intensity of hurricanes gives rise to a need for hurricane classification; that is, for an extension of the Beaufort scale into the hurricane region. This need was met in 1975 by two meteorologists, Herbert Saffir and Robert Simpson, and their Hurricane Damage Scale (see Figure 4-3) is now widely used.

(The unit of pressure used in the scale is the millibar, rather than pounds/square inch. The millibar is one thousandth of a bar, which is the force exerted by 100,000 Newtons over an area of one square meter, where a Newton is the force needed to accelerate a—never mind. The millibar has become the popular unit of measurement because normal atmospheric pressure is nearly one thousand mil-

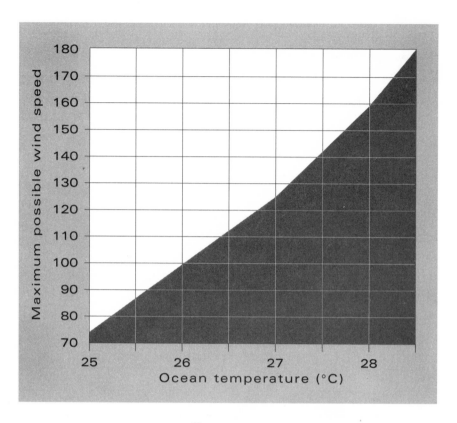

FIGURE 4-2

libars [actually 1013.25 millibars], and so the degree of deviation from this is clearly visualized. My textbook tells me that when we convert to the metric system—"probably early in the 1990s"—this unit will be replaced by the Pascal. Ha!)

The Sunday *Herald* reported Andrew's winds reaching 110 miles per hour as it approached Miami, making it "almost a Category 3 hurricane." That indicates a storm surge of nine to twelve feet, which scared the hell out of me when my father asked if he couldn't stay in his high-rise condo a couple of hundred yards from the Atlantic Ocean.

Hell, no, I said.

CATEGORY	CENTRAL PRESSURE (millibars)	WIND SPEED (mph)	STORM SURGE (ft)	DAMAGE
1	>980	75–95	4–5	Small trees uprooted, mobile homes damaged
2	965–979	96–110	6–8	Roofs damaged
3	945–964	111–130	9–12	Large trees and mobile homes destroyed; structural damage to buildings
4	920–944	131–155	13–18	Extensive structural damage; flooding for several miles inland
5	<920	>155	>18	Whoa, Nellie!

FIGURE 4-3

4

When most people think of hurricanes they think of winds, and it's true that most economic damage is done by the hundred-mile-an-hour-and-up winds, but it's the storm surge that kills people. By the thousands. By the hundreds of thousands.

Take a quick look at the left part of Figure 4.1. The liquid level is at the same height in *a* and *b*. Now imagine that these two spots represent different places in the world ocean. (The U-shaped tube need not be there; it serves only to connect the water at *a* and *b*.) Under normal conditions they assume the same height, but should the air pressure over *a* be reduced while that at *b* stays the same, this will obviously create a rise in water level at *a* (as in the figure on the right).

What could reduce the air pressure over *a*? A hurricane, of course, which begins with a concentrated area of low pressure rising off the ocean surface. (The winds pour into this low-pressure area, it is true, but never bring the pressure back up to normal; if they do, the hurricane is over.) The low air pressure over the growing center of the hurricane acts on the water beneath it just as the evacuated space above a does in Figure 4.1: as the pressure drops, the level of the liquid rises. The most powerful hurricanes have the lowest central pressures—below 920 millibars, according to the Saffir-Simpson scale—and the ocean water at the center can rise more than eighteen feet above normal.

And there are other factors to aggravate the situation. The storm surge is, after all, a sort of combination giant wave and supertide, and so all the factors that normally affect and create winds and tides can do so here as well. All waves are generated by winds blowing over the ocean surface, and the height of the wave depends on how much energy the wind transfers to the water by frictional interaction. The energy transfer depends on three things: the wind speed, the length of time it blows, and the distance that the wind travels over open water. In a hurricane the wind speed is obviously great, but so too are the other two factors, and the resulting waves are much larger than they would be if the hurricane suddenly sprang into existence a short distance offshore.

Another factor affecting the final storm surge is the depth of the ocean floor, which decreases steeply as the storm approaches the continent. This is important because of the behavior of all waves. In deep water the movement of the wave is quite different from that of the water within it. The wave is moving forward, but the water mass is not. A piece of wood floating on the ocean surface remains where it is as successive waves move along beneath it: the wood bobs up and down, marking the movement of the water molecules that surround it: the water molecules move vertically up and down as the wave passes, but they don't move along with it.

This vertical motion diminishes with increased depth beneath the wave. But now what happens when the wave approaches shore and the ocean floor begins to rise toward the continent? When the depth is shallow enough the vertically bobbing water hits bottom, which happens when the depth shortens to about half the wavelength, the distance between successive wave crests. Some of the wave energy is then spent as frictional loss against the ocean bottom—just as the wind transferred energy to the water, now the water transfers energy to the sandy bottom. Other than moving some bottom particles around, this process also causes the wave to lose energy and slow down. And this means that the wave behind it catches up, and merges with it. The resulting wave grows higher and pushes its vertical motion down more forcefully, thereby losing more energy to the ocean floor and slowing further so that succeeding waves catch up and pile into a higher wave. Eventually the total weight of water above sea level is too great to be supported and the wave turns over and crashes down, creating a surf.

This is the normal wave pattern, magnified of course by the increased energy of a hurricane. Normal surf is influenced greatly by the topography of the ocean bottom near the shore—how steeply it rises—and by the distance of open water at great depth over which the waves have traveled. This is why some beaches—at Atlantic City or Ogunquit, Maine—have wonderful surf, while others have little or none. (The greatest disappointment of my life, after realizing at the age of eight or nine that I would never really be able to fly like Superman, was when we moved to Miami and I raced out to the beach with my surfboard and then stood dumbfounded, watching the ocean just sitting there, barely moving. This is because of the Bahamas Bank, which sits offshore and intercepts the ocean waves; the distance between the Bank and the Miami beaches isn't great enough to allow a decent surf to grow, except under exceptionally strong winds.)

Normally, surf is a wonderful thing. But when a hurricane strikes, the combination of excessive wind speed and the long time duration of the storm and the large distance it has traveled means that the waves can rise to a great height before they come crashing down. This effect is magnified with disastrous consequences if the storm hits land at a time of normal high tide. It's a fearsome spectacle to imagine, even more terrible to actually see. When all three factors combine, hurricane waves greater than thirty-feet high can come crashing ashore. Imagine a wall of water as high as a four-story building rushing at you with all the force of a hurricane. For hundreds of thousands of people, this was the last thing they saw on this earth: on average, 90 percent of the loss of life in a hurricane is due to the storm surge washing ashore and flooding the coastal area under a moving mountain of water. The storm surge is the most deadly force in all of nature.

The Pacific is the largest ocean, with, obviously, the greatest chance of waves moving over the greatest distance before hitting shore, and here is indeed where the worst storm surges have hit. The giant tidal waves formed by hurricanes or suboceanic earthquakes, the tsunamis, are part of Japanese lore. But it was on the subcontinent of India that one of the worst disasters in the history of the world occurred.

It happened in Bangladesh, in what was then East Pakistan, on November 13, 1970. Those familiar with the signs knew a storm was brewing in the Bay of Bengal; those who could not read the signs knew nothing, for there was no warning service to speak of. An American weather satellite had spotted a low-pressure area being blown into the bay from the direction of Malay, and had duly notified the Pakistani government. The Pakistani weather bureau followed the storm as it intensified and moved northwest toward the Ganges delta, where a series of densely occupied islands, barely a few feet above sea level, stood waiting for it. The government had built dikes that circled some of these islands, and they were to prove as effectual in stopping the waters as Canute was. The government tried to warn the people what was coming, but there was no effective communication service—no televisions, few radios, and even fewer telephones.

It didn't really matter. Knowing the storm was coming could not have averted it, and there was no way for the people to evacuate: no highways, no cars, no airplanes. Some of the people heard the radio broadcasts and knew that a bad storm was on its way: they boarded up their windows. What else could they do?

Useless, totally useless. About midnight the people on the islands farthest from shore heard a rumbling sound that grew slowly at first but turned quickly into a frightening roar. Peering out their windows they saw a luminous cloud in the darkness, sitting low on the water. As they watched, the cloud came closer, growing in height and roaring until the ground shook and the flimsy wooden houses trembled, and then it rose up out of the waters and they saw that the glittering whiteness was the crest of a wave which towered over them and then came crashing down—and that was the last sight they ever saw.

> *And what rough beast, its hour come round at last,*
> *Slouches toward Bethlehem to be born?*

Later, six children remembered their grandfather gathering them up and throwing them into a wooden crate. Three days passed before the crate was seen bobbing up and down in the ocean, and the children were rescued. But they never saw their grandfather, or any of their family, again.

Reporters who flew over the islands in the days following saw corpses floating in the paddies, corpses littering the beaches, the bodies of men and women mixed with dead cattle and horses. Too many to count, too many to bury. Estimates run close to 1 million people dead, although to this day no one knows how many for sure. "What can we do?" the president of Pakistan explained. "We can only pray to Allah for mercy."

Right. Lots of luck.

5

We didn't expect anything that bad in Miami. For one thing, the ocean floor profile isn't as perfect for generating tidal waves as the one at the Ganges delta. For the same reason there's no decent surf at Miami, the storm surge would be a lot less. Further, the winds of the typhoon that struck Bangladesh were estimated at nearly 150 miles per hour, making it a class 4—almost a class 5—storm, while Andrew was approaching as only a class 3. And it looked as though the surge would hit, if at all, at a time of low tide.

Nonetheless, I still vetoed my father's suggestion that he might stay in his condo, since I was aware of a computer modeling program named SPLASH (Special Program to List Amplitudes of Surge from Hurricanes) developed at the National Weather Service by C. P. Jelesnianski, which simulated the effects of a surge on various parts of the United States coastline. The result for Miami indicated fifteen-foot flooding along the beaches and outlying islands, and five to ten feet for a good distance inland. There was even danger of severe flood damage as far west as I live, ten miles from the coast. Of course, that was for a class 5 storm, and Andrew was only a 3, but even at that class there was a strong possibility of surge waves crashing into Dad's condo. His argument for staying was that he lived on the eleventh floor and therefore need not worry about flooding. I told him that if the surge hit his building, his eleventh floor would soon be the first floor, beneath a fifteen-foot wave. Reluctantly, he agreed to evacuate to my house, twenty miles south and ten miles west.

All over Miami that story was being repeated: many people were reluctant to believe that anything terrible might happen. Initially, the authorities worried that they might not have enough shelters for everyone who wanted to evacuate. Mark Goggin, a Red Cross relief specialist in Broward County (just north of Miami, encompassing Fort Lauderdale), was warning people that the county shelters should be used only as a last resort. The primary option should be for people to get well clear of the area. "We only have room for 110,000," he said, "and that's a tight fit."

As it turned out, there was plenty of room. By late afternoon the high school in North Miami Beach, where Dad lives, was filled with people, but this was the exception. All over Miami the other emergency shelters remained well below capacity. As we ate dinner by the television we were told that twenty thousand people were in various Dade County shelters, which had room for 100,000. The cameras panned around the rooms, and the people smiled and lifted their thumbs or waved for people to join them. The announcers kept reminding us that there was plenty of room, and they kept repeating the authorities' warning that all coastal areas—everything east of U.S. Highway 1—should be evacuated. The inland hotels were all filled, but there was plenty of room in the shelters.

They were giving a party, and no one was coming. I couldn't blame the people who chose to stay at home. Although there was a party atmosphere in the shelters at the moment, it wouldn't last long after the electricity died. Moreover, a hurricane shelter necessarily lacks windows, and when the air-conditioning shut down the air would quickly turn fetid. I didn't imagine the bathroom situation would be comfortable. North Miami Beach High was built to hold a couple of thousand students during the daytime hours; it now held several thousand more people than that, who might be holed up there for several days. In a way, I didn't blame others for wanting to remain in the comfort of their homes.

Their attitude wasn't quite as bad as it had been in Galveston, Texas, on September 7, 1900. For several days rumors of an approaching storm had swirled around. Disembarking sailors had hurried into the nearest bars and whorehouses—nothing strange about that—to tell stories of sailing through the mother of all storms. Well, nothing very strange about that, either. But this time the stories didn't vaporize with the first whiff of intoxicants; it seemed that they really had come through an extraordinary storm. The United States Weather Service wasn't very old then, and people didn't pay much attention to it, but other reports of a serious meteorological disturbance out in the Gulf began to drift into town.

The newspaper editors decided to act. They published a front-page story warning of an approaching hurricane and urging residents to evacuate the city. The next day, due to the power of the press, the population of the city *increased:* no one left, and sightseers came in to see what was happening.

I can understand that. If you've never been through one, a hurricane sounds like *fun.* When I was six years old a hurricane—they called it a nor'easter then—hit Atlantic City, New Jersey, where I was staying with my grandmother. There couldn't have been much warning, but we didn't even get a whiff of one because my grandmother never listened to the radio. All we knew was that the winds began to blow, and from the balcony of her apartment I could see the waves rising up in the ocean. I slipped into my bathing suit and nearly made it out the door before she caught me and dragged me back in.

"Look at the waves!" I implored her.

"Look at the waves!" she warned me, and although we were using the same words, we weren't saying the same thing. I wanted to get out into that surf, and couldn't understand why she wouldn't let me.

All day the winds got stronger and stronger, and I was exhilarated by the growing storm and frustrated at being kept inside. I must have been hell to live with that day, dancing from one room to the other, opening the door to the balcony and running out to feel the breezes, getting hauled back in and sneaking out again. It was *fun;* out there in the ocean, I thought, it must be *heaven.*

Eventually night came and I fell asleep. The next day my grandmother took me out to have a look, and I was literally stunned. The boardwalk, which had seemed to me as permanent as Franklin Roosevelt—they both had been there all my life—was demolished, ripped and torn and broken. I saw it, but I couldn't quite believe it, and I held her hand very tightly.

In Galveston, in 1900, the pre-hurricane euphoria was much the same. They picnicked on the beach and dashed into the raging surf, laughing hysterically at the monstrous waves that picked them up and threw them down onto the soft sand. The weather bureau sent out a man in a horse and carriage to warn them away, but they laughed at him. It was *fun.*

If only the winds had just held that velocity, it would have been great. But inexorably they increased. The blankets people sat on were whipped away if they stood up, and young men ran after them, still laughing. The picnic plates began to sail through the air; chicken and baked beans became messy projectiles. Everyone laughed.

And the winds continued to increase. Children began to cry as the sand flew into their eyes. Deck chairs overturned and blew away, umbrellas turned inside out or flew out of people's hands or out of where they had been anchored in the sand. It began to get too rough, and some began to leave. As the winds increased, spray flew from the surf, soaking the people, and encouraging the exodus.

The first telephone pole went down and people stopped laughing. As they got into their buggies to drive away, tiles began to fly off roofs. These were heavy tiles, installed by order of the city code as a guard against spreading fires, and when they hit they hurt. They more than hurt: people began to fall, and they lay where they had fallen.

Panic took hold. People lashed their horses to get them home, but arrived to find that their homes were no refuge. The waves began to leap from the ocean and wash ashore more quickly than horses could run, overflowing the beaches and filling the streets. Most of the beach homes were small wooden structures that tumbled with the first wave, and the people were suddenly frightened, terrified, lost and alone.

They ran to the large brick and stone houses, banged on the doors, and were taken inside. They were safe there, for the next few hours. But soon water came under doorjambs and broke in through windows. As it came pouring and roaring in, the people climbed the stairs and huddled in the attic until the houses gave way beneath them. And then they died, battered by debris, drowned, suffocated. Fifty people had crowded into the Cline home—a solid stone building—where they died when the solid stone turned into crumbling rubble and collapsed. Hundreds more died in the Cathedral of St. Mary, and nearly as many in the Old Women's Home. In all, some nine thousand people died that day.

The survivors stepped outside the next morning to witness a hellish carnage. Bodies were everywhere: some tangled in trees twenty feet high, drowned up there in the air; dangling from the steel girders of bridges; strewn on the streets; piled in lumps of debris.

Prayers of thanksgiving filled the streets when people realized they had lived through it, but soon these prayers turned into appeals for help from the Almighty as they realized in what condition they had been left. There was no electricity, although this new luxury was not essential to most people yet. But there was no water left unsalted by the sea, no food unruined, no medical supplies, no dry clothing. On top of it all, there was no way out: the railroad tracks were torn up, the bridges to the mainland were down, and the boats had been blown onto the shore and demolished.

The people were on their own, with no hope of help reaching them very quickly. They organized into groups and got to work. One of the most important jobs was to remove the dead before they infected the living. With no time to dig graves, they loaded the corpses onto barges, towed them away, and buried them at sea. At least, people *thought* they were buried there, but in a grotesque joke suitable for the wit of the Marquis de Sade they came home again. Some of the hurriedly organized burial parties were composed of people who didn't know that bacterial reactions in dead bodies lead to the generation of gas; that is, that dead people float. Failing to weight them down, they simply dumped them overboard. Nearly as quickly as the burial barges returned to sea the corpses came sailing after them, floating high in the water, blown landward by the last vestiges of the dying hurricane.

The groups organized by the citizens of Galveston evolved into a city council, which worked so well they decided to keep it as a permanent form of government. From it have descended all the city councils we have in different cities today. Galveston recovered to become a thriving city, an intrinsic part of the Texas economic community. But for another Texas city, things didn't work out so well.

In the latter half of the last century Indianola was Galveston's most intense commercial competitor. Located on the coast just southwest of Galveston, it shared the same advantages and shipping opportunities from the Gulf, as well as the same susceptibility to hurricanes. In 1875 a large one arrived, sending huge waves through the town, blasting it with wind, drowning large numbers of people, and destroying most of the buildings.

But the people fought back by erecting the buildings again, and settling in again. The fledgling United States Weather Service established an outpost there to monitor the Gulf weather. It didn't help; another hurricane hit in 1886. When it was gone, so was Indianola. The weather station was destroyed, and the operator on duty drowned. Enough was enough, the few survivors said, and they searched through the wreckage to find what they could of their possessions, and they left. Today there is nothing to tell the passerby that an ambitious town once stood on the scrub and sand, although if you were to dig deep enough you might uncover the timbers and understructures of what had been homes and businesses, schools and churches; you would also find quite a few bones.

The same thing happened, slightly differently, in the south of Florida. In the early 1920s a land boom was manufactured by smart Yankee real estate operators who took out large newspaper ads that failed to mention Miami's humidity and mosquitoes, luring people in New York and Pennsylvania to buy oceanfront land. My grandfather was one of those who invested. But in 1928 a hurricane stormed ashore and washed the few buildings out to sea, and the boom went bust. Nobody could live in such a place, the Yankee stock market decided, and prices dropped through the floor. My grandfather sold his land for a few pennies on the dollar.

But, as I said, there was a difference between that situation and that of Indianola. Thanks to the inventions of air-conditioning and DDT, as well as the railroad Henry Flagler built down the eastern seaboard to Miami, the land my grandfather sold is today called Forty-fourth Street and Collins Avenue, and the Fontainebleau Hotel sits there.

6

We didn't expect anything as bad as the Texas hurricanes in Miami. But the point was that they hadn't expected anything that bad in Galveston or Indianola, either. Andrew was less than two hundred miles away and still heading due west, right at us. The latest predictions suggested it would pass just below Miami, right into Homestead, a small town about thirty miles south, home to my taxman and the Homestead Air Force Base. This was bad news for me because of the way a hurricane circulates.

Its counterclockwise rotation means that the winds north of center are moving west, while the winds south of center are moving east. If the hurricane itself is moving west, then the effect of the wind speed north of it is increased by the velocity of the hurricane itself, while the wind speeds south of it are diminished by the same amount. This means that the worst damage inflicted by a westward-moving hurricane is to the north of center, and the radius of greatest intensity is about thirty miles, so if the storm kept moving in its present direction it would hit just below us and we would get its brunt.

But I didn't believe that would happen for two reasons. First, as I said, hurricanes always move west *and north* until they curve back out into the Atlantic. Time was getting tight now, as it continued due west, but I was sure it would bend north at the last minute and hit North Miami Beach or Fort Lauderdale. The second and main reason I was sure of my prediction was simple: it could not hit *me.* Someone else, sure, I could accept that; hurricanes kill people. But not me.

Nevertheless, I was glad now that I hadn't bought a house on Key Biscayne, Miami Beach, or anywhere else near the water. There aren't any flimsy wooden shacks on the beaches here like there were at Indianola (houses in Miami are built of solid stone and concrete because of termites), but a solid stone house buried under thirty feet of water is no less a death trap than a wooden house is.

The hurricane that tore up the Atlantic City boardwalk in 1938 hadn't even hit there, it had just churned by offshore on its way north. But the waves it lifted up out of the sea as it passed were enough to smash the city and its famous landmark to smithereens. The hurricane swamped Atlantic City as it passed and slammed into Long Island a few hours later, taking everyone by surprise and raising even worse storm waves there.

Somehow it had evaded the ships at sea, and there was no radar or reconnaissance planes, so few people even suspected its presence until it hit. *The New York Times* was full of stories of potential disaster that day, but its eyes were focused not on the approaching storm but half a world away, on Czechoslovakia: WORLD WAR PERIL, the headlines shouted. HITLER READY TO MARCH. Not a word about the hurricane that was zooming up from just beyond the horizon.

On Long Island it was a beautiful Indian Summer day right up to the last minute. People were tanning, enjoying themselves, splashing in the normally rough Atlantic surf, when the wind began to blow just a bit harder at 2:45 in the afternoon. It looked to most people in the Hamptons like a typical summer squall moving in and, cursing the end of a good day, they gathered up their things and began to leave.

Many of them never got home, and those who did found they were no safer there than elsewhere. Within an hour the waves began to lift up out of the sea as if pulled with hidden strings by a giant invisible hand. They washed over the beach, onto the beach roads. By five o'clock a series of gigantic waves, thirty to forty feet high, rose up and crashed down, riding their momentum far onto the island.

At Westhampton fifty "substantially built" homes were washed into the sea as the waves swept over the dunes and all the way in to Main Street, where the water reached a height of eight feet. The electricity went out all over the island, and the darkness was unrelieved when night fell early with the thick cloud covering. Although there was no electricity on the island, some of the downed high-tension wires were still hot and sizzling in the pools of water; one man, casually brushing a fallen wire from his car, was immediately electrocuted. For "the whole length of the island," reported the *Times*, the towns were dark shambles where trees were blown over, telephone and electrical poles were fallen, and streets were flooded.

An eighteen-year-old girl was killed when a brick wall toppled onto her car, which was stalled in several feet of water. The Southampton Beach Club was completely demolished, and the *Times* reported that "the storm dealt out death indiscriminately. Among the victims were prominent members of the colony as well as Negro servants." (Apparently this nonpartisanship on the part of God was not expected by the *Times*.) Human interest stories followed: "Mrs. Margaret Delahanty, 50 years old, became panic-stricken when huge combers began pounding her house. Her husband, John, put her in his car and made the hazardous crossing to the mainland." Safe on the other side, she was still trembling with fear. Worried about her, he took her directly to a doctor's office. As she entered, she dropped dead.

But the best story was covered by *Time* magazine. Tex Langford, a twenty-three-year-old cowboy, had come put-putting into New York Harbor the previous week in his little outboard. Approaching the pier, he stood looking at a huge yacht moored nearby. She was dirty and her masts were missing, but she was *yar*, all right. As he climbed up on the pier he saw a "benign-looking stranger gazing off to sea," and Tex, making casual conversation as he tied up his small boat, allowed as how that was a mighty fine ship moored out there.

"Glad you like her," said the stranger. "She's yours."

The stranger was John S. Nairns, a maritime inventor who was tired of the sixty-three-foot yacht *Winnetta*, and wanted to give her a good home. That same day he transferred the ownership papers to Langford, and for the rest of the week Tex scrubbed her down, cleaned her out, and got her ready for his life's dream: a cruise around the world by sail.

And then the hurricane hit. The *Winnetta* was found afterward, "jagged driftwood" among the hundreds of other boats broken and shattered. And Tex, like his dream, just faded away.

Aside from human-interest stories there was nothing but unmitigated disaster. In Rhode Island the hurricane hit as school was letting out. A school-bus driver, not realizing the extent of what was happening, picked up a half-dozen children and tried to drive them over the causeway to their homes. One child's father, hurrying from the other side of the causeway to pick up his son at school, had seen that the causeway was flooded and impassable. Stuck there, wondering what to do, he saw the school bus approach from the other side. He tried to wave it off but the driver didn't see him and kept coming, got halfway across, and then the bus was washed off the road into the waters. The man watched as the bus driver managed to get the door open and the children climbed out on to the roof. He saw them holding hands there for a few moments as the bus floated away, and then as it sank he saw the children try to swim for shore. One by one they disappeared; one of them made it to dry land, but not his son.

All in all nearly a thousand people died, several thousand homes were destroyed, hundreds of millions of dollars were lost.

7

It is the storm surge that does most of the damage, but I wasn't personally worried about the surge in Miami as Andrew approached; we were too far inland. Yet it was looking more and more as if Andrew would really hit; if not us, then surely somewhere on the Florida coast. It was still heading due west and was just over a hundred miles away, the television told us—too late now for it to curve away harmlessly to sea. Even if it turned north, as it should, there wasn't enough room left for it to miss the coastline.

It was also speeding up and gathering strength. This morning it had been a category 3 storm, now it was a 4 and still building; there was talk of it reaching category 5 status by the time it hit us.

Category 5 is nearly unheard of. Since records have been kept only two such storms have struck the United States: the great Florida hurricane of 1935 and Hurricane Camille, which hit Louisiana, Mississippi, and Virginia in 1969.

The 1935 storm hit the Keys, just below Miami, with impossibly potent winds. People caught out in the open were sandblasted, quite literally: the sand was lifted up from the beaches and blasted them with such force that it stripped their clothing away and scoured their skin. Corpses were found with nothing left but their leather belts and shoes—and when I say "with nothing" I mean not even their skin remained. A fifteen-foot wooden log was picked up by the wind and thrown right through a house. A meteorologist at the weather station reported that the station house itself was blown down by the wind, and he was picked up and "carried away by the wind out to sea, where I caught hold of coconut leaves to save my life. I was hit by something and lost consciousness. I regained my senses . . . and found that I was stuck in the branches of a palm tree 20 feet above the ground."

When the storm hit down in the Keys, people in Miami had no idea what was happening. They had been expecting the storm, it had begun to rain, and then nothing else happened. The sun came out, the storm was gone. They didn't know where it had gone until the next day, when a few survivors came drifting up in small boats, begging for help. Among the doctors who hurried down was G. C. Franklin, who found a man sitting "calmly against a broken wall with a piece of two-by-four run completely through him, under his ribs, out over the kidneys. He refused the shot of morphine the doctor offered him, before he pulled it out. The man said that when it was pulled out he would die. He asked for two beers, drank them and said, 'Now pull.' Dr. Franklin pulled, and he died."

In the first few weeks of the 1969 hurricane season, two small storms approached the East Coast and then curved northward and passed harmlessly out of sight, doing nothing to attract the nation's attention, which was directed elsewhere. It was a time of great events; Neil Armstrong had just come back from the moon, and our eyes were fixed outward at the stars and planets. And nearly all the nation's children, it seemed, had gathered at a farm in Bethel, New

York, for the Woodstock Music and Art Fair, and our eyes rather nervously turned to them. *Time* magazine called it "History's Biggest Happening," and showed a picture of what looked like boys and girls swimming nude together. *Newsweek* showed a clearer photo, with female breasts actually visible. People wondered aloud what was happening to the world.

What was happening, aside from a coordinated attack by the Vietcong after a two-month pause, was a tropical storm gathering itself together in the Caribbean. On August 16 it blew ashore on the Gulf Coast, smashing and tearing it up from Mobile to New Orleans. It was "the greatest storm of any kind that has ever affected this nation," according to Dr. Robert Simpson, head of the National Hurricane Center in Miami.

In Pass Christian, a small resort town on the Gulf in Mississippi, three thousand people evacuated as the storm approached. Afterward the other two thousand—or what was left of them—wished they had, too. As Camille headed toward them, Police Chief Jerry Peralta rode up and down U.S. 90, the highway that runs along the beachfront, stopping at every house and warning the people to leave. Some did, many did not. At the posh Richelieu Apartments they laughed at him. A group of twenty or thirty young adults were organizing a hurricane party. Most had been through hurricanes before, and it was kind of fun, they said. The electricity would go out and the ocean would turn into a gorgeous wild surf and the wind would blow, and everyone would be stuck there for a couple of days. They were prepared: they had plenty of beer. "One man told me this was his land, and if I wanted him to leave I'd have to arrest him," Peralta later recounted. He couldn't do that, but he insisted on taking the names of the next of kin of each of them. "They laughed at that," he said.

The storm hit at ten that night. The electricity went out almost immediately, which wouldn't have disturbed the party at the Richelieu; indeed, that would have been part of the festivities. But then a wave "as high as a three-story house" rose up out of the sea and crashed over them. That was the end of the party.

The wind demolished nearly all the beachfront buildings and scoured the rest of the town. It tore through Trinity Episcopal Church and sucked four children out of their father's arms and flung them into the dark night, never to be seen alive again. A pickup truck sailed through the air like a kite for fifty yards before crashing to earth. For four hours the storm raged without mercy.

Police Chief Peralta took charge of an amphibious National Guard Duck that arrived on the scene, and went looking for survivors. "I guess we found fifty people. . . . They were hanging in trees and just screaming and crying. At the school, the people inside were holding their babies over their heads. It was all over and the wind had stopped and the tide was going down, but they was just standing there and screaming and holding their kids up."

Of the twenty or so people who had insisted on staying to party in the Richelieu Apartments, none of the adults survived. The solid stone building was shattered by the storm surge, and all its wood and stone and the broken bodies of the people inside were blown across Pass Christian. The next day a five-year-old boy was found floating some distance away on a mattress. He was the only one to live.

8

So a force 5 hurricane was something to worry about, and the people in Dad's condo were crazy to stay there. Even as far inland as we were, you had to be worried, because it was the wind that had shattered Trinity Episcopal in Pass Christian, not the waves. The hurricane would begin to lose strength once its center came over land, but the strongest winds would likely be spinning out fifteen or twenty miles from the center, and my home was about ten miles inland, which meant that the full force of the winds would hit us before the center reached landfall. Even then, its loss of energy would not be instantaneous. All in all, though we were far enough inland to be beyond the storm surge, and though our stone home should withstand any normal hurricane, a force 5 was something to worry about.

And it was coming fast. It had been first expected to hit Tuesday morning; predictions were revised to Monday afternoon, then Monday morning, and finally sometime Sunday night between midnight and dawn.

That evening I stood outside the front door of my house, looking east, waiting. Up and down the street my neighbors were doing the same. Nothing happened. There was nothing to be seen, no wind to be felt, no roaring cataracts of sound to be heard. I thought of the people in London in the winter of 1940, standing outside in the twilight, listening to the silence, waiting for the night to fall and the Heinkels and Dorniers to come. My neighbors and I looked at each other, smiled, shrugged, and went back indoors. To wait.

Unlike the Long Islanders in 1938 or the Texans in 1900, we knew what was coming, thanks to Joseph P. Duckworth . . .

5

STORM HUNTING

Some day I'll find you,
Moonlight behind you . . .

—from "Private Lives,"
by Noel Coward

1

One of the bravest unknown men in history, it is said, is the man who first ate an oyster. An even braver man, I suggest, is Colonel Joseph P. Duckworth, United States Army Air Force, who on July 27, 1943, first flew an airplane into a hurricane.

Duckworth had won his wings with the army in 1928 and had later flown as a captain for Eastern Airlines. In 1940, with war approaching, he took up his commission in the reserves and was assigned as a flying instructor. At that time, he found, the army knew nothing about instrument flying. Up until a few years previous, the airlines had known just as little; for example, the way they found their course at night was to look for lights from cities, and if there was no city in view they searched for the lights from cities beyond the horizon reflected on the underside of clouds; comparing these reflections with their maps, they found their way.

Another use of reflected lights was to monitor their intensity: when they grew brighter it meant the clouds were dropping, and pilots were instructed to get down at the nearest airport as quickly as possible to avoid getting caught in the haze.

Flying without visibility in those days was asking for an early death. Years later, as I was flying a small Piper Cub from northern

Vermont down to Hartford, I was caught in similar conditions. Although instruments were in general use, I was only qualified to operate a plane under VFR (visual flight rule) conditions. A low mountain range was in front of me, and high cloud cover was above. No problem, I thought, except for the fact that the clouds kept lowering.

I should have turned and gone back to Vermont to fly another day, but I was young and impetuous and stupid. I kept lowering my altitude to stay under the haze until finally I saw the mountains ahead reaching into the clouds: their tops weren't visible. I hauled around on the stick to turn then, but it was too late. As I circled I saw that the haze behind me was even lower than the stuff in front. I had two alternatives: climb up through the crud and find clear skies above or set the plane down right away in the nearest field.

I should have set her down. I still don't know why I climbed instead. It was probably because I was still trying to convince my wife that flying a light plane was a safe means of transportation, and I didn't want to have to call her from a farmhouse to say I had made an emergency landing and wouldn't be home that day. If I did that, I'd never get her up in a plane again.

Somehow I got up through the muck without getting discombobulated, or tipping a wing over and spinning out, and I flew ahead in brilliant sunshine with the white carpet unbroken below me. I was hoping that it would clear in a little while (my Cub had no radio in it, so I didn't know what conditions existed ahead. Stupid, stupid). Eventually I was sure I had passed over the mountains and had to be in the Connecticut Valley region, but there were still no breaks in the clouds. I circled around looking for an opening, but all I could see was my fuel level getting lower and lower. If the damned plane could have flown on perspiration, I would have had no problem staying up there for a couple of weeks.

But it couldn't and the fuel continued to diminish, and finally I had no choice but to go down while I still had power. I dipped into the white stuff and was immediately blind, confused, claustrophobic, and paranoiac. I concentrated on my only three instruments: I kept the air speed as low as I could without stalling so that if a mountaintop suddenly reared up in front of me, I might have a moment to avoid it; I kept the ball-and-bubble centered to keep the wings level so I wouldn't spin down into the ground; and I watched the altimeter unwind toward zero.

I got lower than the mountains and still hadn't crashed, so I must have been right about passing them. But I didn't feel very good about that since the altimeter kept unwinding; pretty soon I would be below ground level although I was still surrounded by clouds.

And then, at a couple of hundred feet I broke out suddenly into the clear, and everything was all right again. I got home on time and never mentioned a word about this to my wife.

Army air force pilots in the early 1940s weren't much better than I at instrument flying, and with a war approaching that was a serious problem. Duckworth was assigned to rectify the situation, and he set up an instrument flying school at Bryan, Texas, where on July 27, 1943, he heard of a hurricane blowing around in the Gulf.

The regional forecaster in Louisiana, a man named Stevens, had suspected the presence of a tropical storm from the routinely meager data available to him. Owing to the fact that German submarines were stalking our shipping along the coasts and across the ocean, all ships at sea kept radio silence for fear of detection, and thus provided no weather information to the mainland. The only data Stevens had to work with were reports from coastal weather stations, yet he was able not only to guess at the presence of a storm but to predict it would grow to hurricane strength.

Early in the morning the storm blew inland between Bryan and Galveston, and Duckworth found out about it. He grabbed a young navigator who didn't have the sense he was born with, Lieutenant Ralph O'Hair, and together they set out in a single-engine trainer, a North American AT-6, to take a look.

Stupid, stupid. They didn't have a clue how strong the storm might be—the forecaster didn't know if it was a force 1 or a force 5 hurricane—and it was impossible to know what a hurricane might do to a plane. They had considered the turbulence, of course, but had neglected another factor that proved important in later flights: rain.

Duckworth never worried about it because he had flown through rainstorms routinely in the past few years. But in a large hurricane the amount of water evaporated from the ocean is absolutely tremendous, so much so that flying through the storm can be nearly like flying underwater—which can't be done. Although Duckworth had only one engine in his AT-6, luckily for him the storm was a small one and had moved inland near Galveston before he caught up to it. He flew in and back out again without any trouble. In fact, when he got back to base and the local weatherman said he wished he had been along, Duckworth popped him into the AT-6 backseat and took off again.

In subsequent flights, aviators found their engines sometimes conking out in the storm; some brought their planes back with one or even two engines out, and others never came back at all. One of the worst flights was accidental: in 1947 a Pan Am (then Pan American) flight from Cuba to Miami inadvertently caught up with a hurricane and flew right into it. Aboard was Eddie Jones, an Associated Press reporter:

> Through some fault in weather information, the crew left Havana believing they would follow the storm into Miami. Instead, the plane virtually passed through the eye of the vicious disturbance. . . . Seven passengers passed out cold from fright. Others dripped cold sweat, . . . Mrs. Jones and I can testify to the cold sweat.
>
> The group left Havana at 5:45 P.M. Saturday in a torrential downpour and sped out over the Florida Straits. They witnessed a colorful sunset above the clouds—and nosed into the storm. The big airplane was tossed about like a feather. Vertical air blasts sometimes threw it straight up 1,000 feet or more, sometimes dropped it sickeningly as great a distance. Lightning, brilliant and blinding, flickered around the plane constantly and once a ball of St. Elmo's fire leaped across the wing.
>
> Rain in such torrents that only the nearest of engines could be seen, hammered against the fuselage, flooded into the pilot's

cockpit, and dripped down the cabin walls. Twice the plane ran through hail and the sound was like the wings being torn off.

The wildly bucking and weaving plane twisted so that the cabin door leading out of the plane flew open. Steward F. S. Yado unbuckled his safety belt and, held by Stewardess A. A. Brady, caught the door and pulled it shut again. So wild was the ride that to unbuckle one's safety belt was to invite a tossing from end to end of the cabin.

The crew finally found Miami, but the airport was closed. Landing in the storm was made impossible by the wind and lack of visibility. After trying for an hour and a half they flew to Nassau, where the crew got a standing ovation after setting the plane down. The next morning they flew back to Miami, and were impeded by a flickering indicator light for one of the wheels. They circled for an hour, using up fuel to prevent an explosion during a belly landing, and finally found that the wheel was down after all and they were able to land safely.

By this time the army and navy pilots had established routine aerial surveillance of hurricanes, which meant not only finding the storms and tracking their progress but flying right into the center. This was the beginning of our modern understanding of hurricanes, and like the very first beginning, which centered around The Great Hurricane of 1780, it was the result of a war.

In the early 1940s, while the United States was defending itself against Germany, there was an unfounded but very real fear that Hitler would launch an offensive against the American continent by invading the West Indies. It was unfounded because Hitler didn't have any aircraft carriers, and the Battle of Britain in 1940 and subsequent naval actions in the Pacific were showing that you cannot launch an invasion without adequate air cover. But at the time there was little confidence that our land-based bombers would be able to repulse the German navy if it came.

The threat receded as we invaded Africa and the Russians broke the German Army at Stalingrad, but before the significance of those events sunk in we had decided weather information was needed throughout the Indies year-round. In particular, in order to predict the evolution of a burgeoning hurricane, weathermen had to know its central pressure, since that is what determines wind growth.

The first flights brought a spate of new knowledge. For one thing they found that the most violent winds were down low, while up high the terrible updrafts and turbulence subsided into a more orderly organization. It's not the speed of the winds that's dangerous to an airplane, since the plane is moving within the wind. Flying into a two-hundred-mile-an-hour wind presents no problem if the wind is steady. You simply move along with it like a swimmer in a smooth current, marveling at how fast you're going relative to the shore, if you're going along with it, or how slow if you're trying to swim against it. But it's easy enough to swim—or fly—in the current. Planes routinely take advantage of the two-hundred-mile-an-hour jet streams on flights to Europe. But when the water or air gets turbulent, when the waves come crashing down or the updrafts suck you upward or downdrafts force you downward, that's when it gets hairy.

This explains why hurricane winds are more violent at low altitudes: it is there that winds come into the central low-pressure area and suddenly rise into the vacuum. They spiral inward from all directions, rising up laden with water vapor, mixing and roiling terribly. At high altitudes all this commotion gets relatively sorted out, and although the wind speeds are high they are more continuous and easier to fly through. People didn't understand the physics behind these observations in the 1940s, but the observation itself was enough: when flying into a hurricane the recommended procedure was to approach it at high altitude and gradually sink down; if the weather got too rough, a pilot could simply climb out of it again.

That made sense from the pilot's point of view, but not from the navy's. The navy needed information taken at sea level, where the violence of a hurricane is greatest, because, after all, that is where the navy's ships are. Flying at a high altitude might be safer, but it couldn't determine what was happening at sea level. So the navy adopted a different strategy, trying to sneak in *under* the clouds.

This didn't enable them to avoid the violence of the winds, but at least staying under the clouds permitted them to see how high they were above the sea. In those days altitude in an airplane was measured with a barometer, which simply measured the air pressure outside the airplane. Since the pressure under normal conditions decreases at a steady rate with increasing altitude, the dial of the barometer/altimeter could be calibrated to give the height above sea level. But note the words *under normal conditions.* In the center of the hurricane the pressure drops far below normal. So if you're flying in a thick cloud one hundred meters above the ocean, under normal conditions the instrument would read an outside pressure of about one thousand millibars. But in a hurricane the pressure might drop 10 percent below normal, and so the altimeter would read only nine hundred millibars, suggesting an altitude of one *thousand* meters. If you dropped out of a cloud thinking you had a thousand meters of cushioning air below you in which to recover, and found instead the ocean waves reaching up for you only a hundred meters away, those oceanic fingers would probably be the last sight you would see on this earth.

Today planes are equipped with radar altimeters. The instrument beams a radio wave at the ground (or water) to measure how long it takes for the beam to reach the ground and bounce back up, then translates this time into altitude; these altimeters aren't influenced by changes in air pressure. But in those early days of low-level intrusions, pilots had to avoid getting sucked into the clouds at any cost; once inside, with only an unreliable barometric altimeter, there was no way of knowing exactly where the ocean was.

Unfortunately, it wasn't always possible to stay low and clear, especially if the storm had ideas of its own. In 1948 a navy pilot named Captain Desandro reported: "We hit heavy rain and suddenly the airspeed and rate of climb began to increase alarmingly and reached . . . 260 miles per hour and four thousand feet per minute." They were climbing rapidly, and nothing he could do would slow them down. He was afraid to cut the throttle since a sudden wind change could, without warning, find the airplane below flying speed. Stalling in those conditions would have meant dropping into the waters without hesitation. In fact, a few seconds later, the plane did lurch into another wind, and the speed instantaneously dropped more than a hundred miles an hour. They were now flying at 130 miles an hour and dropping like an elevator. "The engineer . . . came up off the floor like he was floating in the air," said Desandro, but luckily they had enough altitude to straighten out before they reached the water.

Flying straight into the waves wasn't the only danger, of course. The sheer strength of the turbulent winds was sometimes greater than the sheer strength of the airplane or the physical strength of the crew. Another pilot described his experience:

> We were shaken by turbulence so severe that it took both pilots to keep the airplane in an upright attitude. At times the updrafts and downdrafts were so severe that I was forced down in my seat so hard I could not lift my head and I could not see the instruments. Other times I was thrown against my safety belt so hard that my arms and legs were of no use momentarily, and I was unable to exert pressure on the controls. . . . We were unable to maintain control of the altitude; all we could do was to hold the airspeed within limits to keep the airplane from tearing up from too much speed or from stalling out from too little. The third pilot . . . was thrown all over the flight deck. . . . To some it may sound exaggerated and utterly fantastic, but to me it was a fight for life.

Upon returning to base, planes were often found to have severe damage: bent and warped tailfins, flaps or ailerons or mountings torn loose, snapped rivets, instruments ripped out of the panel and smashed, sometimes against the pilot's face. But you wouldn't know all this from the Navy's official instructions:

> The storm area is approached on a track leading directly to the storm center and may be approached from any direction. As the

winds increase in velocity, corrections will be made so that the wind is from the left and perpendicular to the track. The point at which the box is started is the midpoint of the base side of the rectangular pattern to be flown around the storm. When winds of sixty knots are encountered, the first leg will be started with a 90° turn to the right.

The low-level box will be flown within the 45–60 knot wind area maintaining a true tack for the first half of the leg, then a true heading for the succeeding legs. Surface winds should be 45° from the right when the left turn is made to the next leg. Double driftwinds should be obtained on each corner observation and each midpoint when practical. Flight altitude while boxing the storm will be a minimum of five hundred feet absolute altitude, or at such higher altitude as will permit observations of the sea surface without hazard to safety. . . .

That was the principle, at any rate. It didn't always work: the rain was often heavy enough to swamp an engine; or the stress was more than the plane could withstand; or the pilots didn't react quickly enough to the tempestuous winds . . . More than one hurricane hunter never came home again.

But from those who did we began to learn how a hurricane works. We found, for example, that the central eye is a complex area, creating warmer temperatures at higher altitudes. That was unexpected, but we understand it now: the storm gets its energy from evaporating ocean water, which in essence draws in energy and releases it by condensing the water vapor into rain. This occurs as the warm winds rise and are unable to hold their moisture, and so it follows that as they rise they get even warmer. This was our first clue to the energy development of the hurricane, the first hint as to how the hurricane actually works.

Some of the first flights into the hurricane eye showed another surprise: strong and steady downdrafts were sometimes encountered, which meant that in the center of the eye the air was moving downward. And measurements taken at ground level by weather stations under passing hurricanes had shown since the beginning of this century that as the pressure fell to a minimum, thus indicating the very center of the storm, the temperature rose to a maximum:

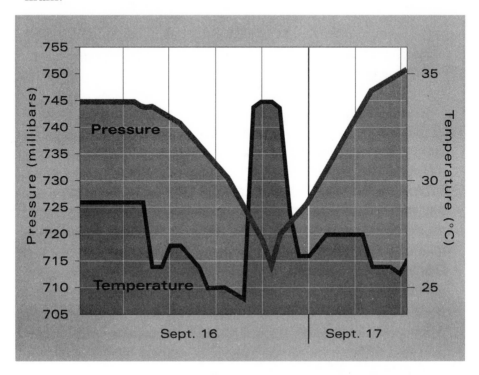

FIGURE 5-1

American meteorologist Bernard Haurwitz postulated in 1935 that the combination of warming with descending air observed in the eye is a natural one, due to descending currents of air that have lost their water vapor during their previous uplifting. If this is so, the air should be not only warm but dry. He tested this by getting data from the Manila typhoon of 1882: "The relative humidity, which was 100 percent before and after the calm (that is, outside the central eye), dropped to 49.7 percent," in agreement with his prediction. He went on to say that

> the rise in temperature might be due to insolation [direct heating of the sun since there are no clouds in the eye] . . . but if the

change in humidity had been caused by insolation . . . we should expect to observe a drop [in humidity] to about 61 percent; but the air in the center is very much drier, indicating that in the eye of the storm we have a downward current which brings warm and dry air to the ground. This air in the center probably comes from the surrounding regions of the cyclone, and has lost part of its moisture content by precipitation during [its] previous ascent.

Ivan Tannehill, reporting on this work, was like other scientists of his day, partly convinced and totally astounded by this conclusion: "There seems to be no other satisfactory solution," he admits. "Yet, if true, it is a paradox. . . ."

The paradox arises because of our model of a hurricane: surface winds stream across the ocean surface to the hurricane's center, which is a region of low pressure, and their ascent and loss of accumulated moisture fuels the storm. If they are rising in the center, how can air be sinking there?

The paradox is solved when we realize that our model is just an idealized approximation. In the real world a hurricane is more complicated, and it is in the dissection of these complications that we learn exactly what a hurricane is, how it will behave, and how we might modify it to our own purposes. We still don't know enough to attain that final objective, or even to predict its future behavior with much certainty more than a few hours in advance, but we are beginning to understand a bit more than we used to.

The hurricane is not a precise instrument put together by Swiss watchmakers, but is rather more like a mélange of disparate parts thrown together by sweatshop workers in Taiwan. The swirling winds are indeed sucked in toward the center by a vortex of low pressure, but they don't get all the way there. Rather, they begin to rise in a series of thunderstorms forming around the wall of the eye. As the air reaches high altitude, having lost its moisture to fuel the storm, *most* of it gets spun away to great distances by the high-altitude anticyclone. But not many things in nature are 100 percent efficient, and neither is this process; that is, some of the dry air pours inward, into the eye, where it drops toward the surface. As it drops it is warmed by compression and, having already lost its water, forms the warm, dry air mass mentioned above.

But doesn't this air, being compressed, form a high-pressure mass? And if so, wouldn't it destroy the low-pressure vortex that is responsible for the hurricane in the first place? The answer is clear: yes and no, perhaps, and sometimes. The hurricane is not a permanent institution; it lasts as long as all the complex mechanisms are working in its favor. If something disturbs it—and a central mass of high-pressure air would certainly do that—it can be disrupted and disorganized, and can disappear. Indeed, all hurricanes eventually do just that. If too much air spills over into the eye too soon, the developing storm might never mature into a hurricane, instead dissipating into a series of weakening thunderstorms. If a small amount of air spills over, however, the storm can handle that and continue boiling away. The answer has to be that a hurricane is not a neat package designed with our understanding in mind; it swirls and whirls, roils and rumbles, and we have to follow it as best we can.

2

All science is like this: the first answers to the basic questions give rise to more questions. Little by little, question by question, confounding confusion and exhaustion, we learn. The first advances in understanding these storms came with the invention of the telegraph and the establishment of national weather services in several countries. France was the first, as mentioned in Chapter 3, because of what happened at Balaclava and some previous skirmishes in the Black Sea.

In 1853 Russia moved her Black Sea fleet against Turkey. When the Russian Navy scored a signal success over the Turks that summer, England and France started to get nervous. Turkey was a chained animal bestriding the trade routes to the east; she maintained order and took orders—she was no threat. But Russia was another story. A large part of nineteenth-century history—and, indeed, that of the twentieth—centers around Russia's efforts to break out into the Mediterranean, the *mare nostrum* not only of Italy but even more so of France and England.

When her one European ally, Austria, withdrew support, Russia gave up her plans of immediate conquest of the Turkish Empire. But England and France had been forewarned of her intentions, and by the next year they were ready to act. They sent a combined force to invade the Crimea, which resulted in some fine jingoistic poetry ("The Charge of the Light Brigade"), the beginning of military nursing (Florence Nightingale), and not much else before the winter of 1854 set in. The invading forces settled down around their supply base at Balaclava to wait for spring.

Unfortunately, before spring came, something else did: a winter storm blew across Europe from the west. It was an ordinary storm until it reached the Black Sea, where it began to suck up moisture and grow. It burst upon the huddled armies at Balaclava without warning and destroyed them. The tents and temporary shacks in which the armies lived were blown away, as the ships at anchor were thrown against one another, broken apart, and sunk. The food and medical supplies on which the army had depended for survival through the bitter winter, although never plentiful, were now at the bottom of the Bosporan waters. The men in the field grew weak and sick; those in hospitals died.

The troops had been encamped on a high plateau, exposed to the wind and cold, cut off from the harbor, when the storm destroyed the road; some horses got through, but for wagons the passage up was impossible. Exposure, disease, and lack of food, clothing, and medicine all combined to annihilate the army before spring arrived.

This was actually nothing new, happening rather routinely to armies in the nineteenth as well as in all previous centuries. The difference was that for the first time newspaper correspondents were on the scene in force, and their dispatches home told the public the full story. No longer portrayed as a mythical thin red line of heroes, the army was now seen to be composed of starving, suffering men.

Napoleon III responded. He called in Urbain Leverrier, the astronomer who had discovered Neptune, and asked him what could be done to prevent any reoccurrence of the storm-thrust tragedy. Leverrier gathered weather reports from across Europe for the two weeks in November preceding the Balaclava storm, and showed that had such information been in hand, the storm could have been predicted. Furthermore, he said, with the new invention of the telegraph, such information could indeed be gathered in time to make useful predictions, in time to take useful precautions. Napoleon accepted and approved the report the same day he received it, and the French weather service was established. England soon followed, and the United States Congress established our own weather service some fifteen years later.

Progress continued for a time, with stations along the American coast reporting by telegraph to Washington. Of course, information about storms developing over the oceans had to come from ships at sea, and they often got to shore barely ahead of the hurricanes that had chased them there. Then, in 1887 in Germany, Heinrich Hertz discovered radio waves. By the end of the century their properties were understood well enough for a young Italian, Guglielmo Marconi, to suggest the possibility of replacing Morse's telegraph wires with these invisible rays of electromagnetism, thus to send "telegraphic" electrical messages via "wireless."

He had more than the idea, he had the passion needed to spend long days and nights working in the laboratory, and the technical skill needed to construct apparatuses for sending and receiving the electromagnetic radiation waves—radio waves, for short—and within a year he was communicating over distances in excess of a mile. But the Italian government was not interested, so he came to England in 1896 and demonstrated his results to post office officials there. When he succeeded in sending a message across the Bristol Channel, local newspapers hailed him as the inventor of the century, and reports drifted back to Italy. The next year he returned to Spezia, where he set up a transmitting station and installed receivers in ships of the Italian navy, enabling them to communicate at distances of up to twelve miles. In front of King Humbert and the Italian Chamber of Deputies, he successfully demonstrated the ability to talk to ships at sea. At that same time, he was setting up stations along the English coast for shore-to-ship transmissions. The very next year the East Goodwin lightship was run down by a steamer; the captain radioed for help to the South Forland lighthouse, and lifeboats were dispatched in time to save the men, who otherwise would have drowned, as sailors had from time immemorial.

In 1912 the *Titanic* struck an iceberg and went down, and for the first time in history the world was able to keep abreast of a catastrophe as it was happening, whetting an appetite that it seems will never be satiated. The wireless-radio operator in that doomed ship stayed at his post, tapping out his SOS, giving exact position and details, bringing rescue vessels to the spot, before he finally sank beneath the waves. The survivors of that disaster owed their lives to him, and to Marconi's radio. From that moment on the radio became one of the necessary instruments aboard any respectable ship.

At first this was a boon to hurricane reporting, as ships at sea could now notify the weather service as soon as they encountered a hurricane. But events soon took a curious turn. Looking back, it seems obvious, although at the time no one realized what was happening.

What was happening was this: a ship would report a hurricane at sea and the weather service would immediately notify all other ships in the vicinity, including those whose travel plans would have taken them into the hurricane's path. These would all, unsurprisingly, change course and avoid the hurricane. The result was that the number of ships at sea lost to hurricanes diminished. But the hurricanes were lost. No one now knew where they were, because all the ships in the area vanished as quick as cats hearing a motorcycle roar. The next thing anyone knew about the storms was when they showed up on the coastline and barged into someone's city.

People got upset. In 1932 several hurricanes were reported by ships in the Gulf of Mexico, and then were lost as the ships avoided the affected areas. None of the storms subsequently hit shore, but between the initial sighting and the time when they were regarded as having vanished, the residents of the coastal Gulf towns didn't know what precautions to take. They were upset because things had changed. It used to be that hurricanes appeared out of nowhere, with no warning. Bad situation. Then, with the advent of radio, ships at sea reported the hurricanes, and people ashore could take precautions. Much better situation. But now the ships were being warned out of harm's way, which was good for them but resulted in a double blast for the people ashore: they were told every time a hurricane was out there, and then never got any further news. So a dozen times a year they had to take precautions, board up their houses, leave town or stock up on toilet paper and groceries, and they were getting sick of it.

The weather service shrugged. What could they do? Groups of citizens responded: "Send out the Coast Guard. What do we pay taxes for?" The weather service responded that it couldn't possibly send ships purposely into hurricane waters; it was too dangerous.

The people mumbled and muttered and went home, and in 1933 there were twenty-one hurricanes reported (or tropical depressions serious enough to raise concern), and the people went to Washington. This was the time of FDR and the New Deal, and for the first time in a long time ordinary citizens felt they had a government that would listen to them. The government did listen, but didn't act: Congress passed legislation authorizing Coast Guard interdiction of hurricanes, but FDR vetoed it on the advice of his naval men; the technology simply was not available, it was too dangerous, the ships would be lost. Similarly today groups of people shout that the government should come up with a cure for AIDS; great idea, but we simply cannot do it yet.

The years puttered by while the situation got no better and people's nerves got worse as hurricanes were repeatedly reported and then lost sight of. By 1937 a grassroots movement had spawned congressional committees and later a joint meeting between the citizens' representatives, some members of Congress, the newly appointed commandant of the Coast Guard—Admiral Waesche—and James Roosevelt (who was standing in for his father). FDR told him to listen to everyone, give them a chance to get things off their chest, and finally explain that naval technology was not yet up to the task of taking on a hurricane. "Make it clear that I will veto the bill again," he said.

So the participants gathered at the White House, and James opened the meeting with a few words of welcome and then called on Admiral Waesche to say a few words. Unbelievably, neither James nor his father had conferred with the admiral beforehand, and James was soon amazed to hear Waesche say that he had been talking to some of the participants before the meeting was called to order, and he knew what they wanted and he thought it was a fine idea. He was prepared to order the Coast Guard out at the first sign of a hurricane, to penetrate it and note its characteristics, and to track it and warn everyone in its path.

He sat down, and everyone looked around, stunned. There was nothing else to say. The conference was over, and rather lamely James Roosevelt banged the gavel and everyone went home. Except for James, who had to go upstairs and explain to his father what had happened.

Well, FDR had to follow through. The National Weather Service received instructions to notify the Coast Guard when it received word of a burgeoning hurricane, and true to his word Admiral Waesche sent out a cutter each time the weathermen called. But the scheme never worked. Once a ship leaves port it is under the supreme command of its captain, and no matter what his sailing orders are, his prime responsibility is to bring the ship safely home again. As Nimitz would later write after the Halsey affair, "The time for taking all measures for a ship's safety is while still able to do so. Nothing is more dangerous than for a seaman to be grudging in taking precautions. . . ."

The Coast Guard captains took the proper precautions: when they saw a hurricane they turned around and sailed right back to port. After a couple of years the program was abandoned.

3

During the WWII years hurricane tracking ran into a double-barreled conflict. When a storm rolled in, aircraft would have to roll out: entire squadrons were evacuated to safer areas to avoid being caught on the ground and destroyed. (This has not changed; the most modern fighter planes in our air force, the F-16s, when caught on the ground by Andrew were destroyed; Homestead Air Force Base was littered with the wreckage.) In peacetime the warning would go to the civilian sector as well as the military, of course; people as well as planes would have to evacuate. But weather knowledge was a military secret during wartime, and rightfully so. Nazi U-boats at the time were engaged in one of the most destructive sieges of what came to be called the Battle of the Atlantic. They were patrolling American coastlines and sinking our ships sailing from New Orleans to New York, from Pensacola to Boston. The one available weapon effective against the subs was air patrol. If the subs heard our radio broadcasts about approaching storms, they would know that our aircraft were being taken away from that area. And of all ships at sea, the submarines were least vulnerable to hurricane damage; they could ride out the worst of the storm at depth, below the roiling waves, and come up to do their damage before the evacuated air force could return.

On the other hand, if cities were not warned of approaching hurricanes, the damage wreaked could have been worse than the submarine fleet could impose. It was a dilemma the military wrestled with throughout the war, coming down sometimes on one side of the argument, sometimes on the other.

It was in the years after the war that the major advances in understanding these storms were made, when as a nation we found that we had a large number of sturdy airplanes and oceangoing ships without an enemy to fight. Some were mothballed, others destroyed, and a large number were kept available and operational for future wars, but there were still some left over. With great good sense the leftovers were used to explore the oceans and the atmosphere, for scientists understood enough to understand that we understood virtually nothing. As late as 1956 Ivan Tannehill had to admit that the basic story of the hurricane was "one of the great mysteries of the sea. Its heated surface lets loose great quantities of moisture which somehow feed the monster—that we know—but what sets it off is almost as much of a mystery as it was in the time of Columbus."

Scientists still don't know everything about these storms, but an awful lot has been learned in the intervening forty years. Much of that knowledge has come from fieldwork, from the flights directly into the storms, and much has come from subsequent laboratory experiments and computer analysis. Take the question of the structure of the eye of the hurricane: the first models indicated this to be a region of uprising air, but Haurwitz's data and some of the first airplane flights indicated downdrafts. The first definitive answer came from a totally new technique, due in its essence to yet another war: the U.S.-Soviet Cold War of the 1960s.

4

The story begins early in 1960, when Dr. Gote Ostlund came to Miami from Sweden to help get the marine laboratory started in the field of modern geology. The lab had begun during World War II when a young British biologist, F. G. Walton Smith, who had been working on sponge-fishery problems for the Colonial Office in the Bahamas, happened to meet Dr. Bowman F. Ashe, the first president of the University of Miami. They got to talking and decided that Miami was the perfect place for America's first institute for tropical marine biology.

By 1960 the marine lab was becoming one of the three or four best oceanographic institutions in the country, and had branched out into geology. Dr. Ostlund had recently built a radiocarbon-dating laboratory in Sweden, and was asked to come set up another in Miami. He did, briefly returned to Sweden, and resettled in Miami for good: "I got sand in my shoes, as they say."

The radiocarbon-dating facility he built in Miami still pumps out thousands of bits of data every year. The counter consists of a long glass tube within which the radiocarbon is deposited as carbon-dioxide gas. Radioactive emissions from the carbon-14 are counted, and extraneous emissions from other (outside) sources are swept away by an enormous amount of metal shielding and by a complex array of electronics. It's a wonderful setup, but even while he was building it Gote Ostlund was interested in another problem he could solve with the same apparatus. Because his counters would count not only the decay of carbon-14 but of any radioactive atom that could be pumped in—like tritium, for example.

When America and Russia detonated hydrogen bombs over Central Asia and the Pacific in a huge series of tests during 1961–62, they thought it was their own business. If the Russians wanted to blow holes in their own territory, or if we wanted to vaporize a (practically) uninhabited island in the vastness of the Pacific, why should it concern anyone else? (Of course there were some local inhabitants who would be inconvenienced, but they were few and far-away and virtually invisible to Americans at home.) What the governments didn't realize—although the scientific community warned them—was that the effects of the bomb would be felt all over the world. The energy they were releasing was too great to be confined within the man-made limits of territorial statehood.

Ordinarily, when a bomb explodes, the debris falls close by. If you make a bigger bomb, the debris gets thrown farther away but it still falls within the general vicinity of the explosion. But with hydrogen bombs something new was happening: the force of the explosion literally blasted a hole in the sky, and the debris was thrown into the stratosphere—where the atmospheric ball game is played by different rules.

We can visualize the troposphere, the lower part of the atmosphere (the bottom five miles or so), as a region of general circulation.

Winds blow both horizontally and vertically. In particular, warm, moist surface air rises, cools, and loses its water vapor as rain. The resulting air then turns over and sinks down again to the surface.

But the stratosphere, like Proust's past, is a different country: they do things differently there. The circulation is horizontal rather than vertical, and the water vapor doesn't rise that high. The result is that there are very strong winds, such as the jet streams, which can carry airliners to Europe with an additional couple of hundred miles per hour under their wings, but there is no real weather: no clouds, no rain.

Therefore, if you set off a bomb blast and throw particles up into the air, or if a volcano erupts and spouts dust skyward, normally not enough energy is involved to get the stuff through the tropopause. So small particles float around in the troposphere for a while because of their large surface-area-to-weight ratio, like leaves slowly fluttering in the wind, but are rather quickly brought back to earth by currents of descending air and especially by rain, which literally washes them out of the sky. But when the theoreticians calculated the force of the hydrogen bombs they were planning to test, they realized it was so great it would literally tear a hole in the troposphere and throw radioactive products right into the stratosphere. We're not talking big boulders here: any dirt or other matter would be vaporized and then condensed into tiny packets of dust. In that state, without the rain and vertical circulation of the lower atmosphere, the stuff could remain aloft a long time. How long? It wasn't clear, but clearly long enough for the jet streams to disperse it around the world before gravity finally sucked it down low enough to be caught by the weather patterns of the troposphere.

Somehow, incredibly, no one listened to the theoreticians, just as no one listened to the cries of local inhabitants in Siberia and Polynesia. The tests went off as planned, and the radioactive debris circled the planet. When it finally settled it brought a new awareness of the term *global community.* Radioactive iodine and strontium settled in the fields of the United States, Russia, and Europe; it was eaten by cows in Ohio and New Jersey, in Bavaria and the Ukraine, in China and Japan, and later showed up in their milk. More than one physicist gave up his job and left for Australia or New Zealand, hoping the stuff wouldn't follow him across the equator (since the jet streams are primarily east-west rather than north-south). This radioactive crud from the bomb tests, together with the pesticide alarm sounded by Rachel Carson in her book *Silent Spring,* effectively started the worldwide environmental movement.

So in a way it was good, perhaps even necessary, because environmental awareness is slow to develop; perhaps if we hadn't got started in the mid-sixties we wouldn't be ready for the much more terrible ozone and greenhouse problems we're facing today.

An additional benefit was that it provided a way to trace the trail of water vapor through the atmosphere, and thus a way to study the developmental structure of hurricanes. One of the products of the hydrogen bomb is tritium, the radioactive isotope of hydrogen that interested Gote Ostlund. He was curious about it because, despite the fact that it is radioactive and heavier than normal hydrogen, neither of these characteristics interferes with its basic chemical behavior as hydrogen. In particular tritium can take the place of one of the two hydrogen atoms in a water molecule. The resulting radioactive molecule is known as tritiated water (sometimes written as HTO instead of HHO or, more commonly, H_2O), but it is still water and behaves like any other water molecule. (There are some subtle differences because it weighs more than a normal molecule, but this does not affect the following discussion.)

The tritium blasted into the stratosphere by hydrogen bombs eventually — over a time period of a few years — finds its way into a water molecule and wanders down into the troposphere, where it takes its place in the global cycling of water between liquid form in the oceans and the vapor phase in the atmosphere: it is removed as rain and falls into the oceans. This transition happens relatively quickly.

The ocean, like the atmosphere, can be broken up into two major areas: a warm surface layer, and the colder deep waters below. The waters in the surface layer mix well and quickly; underneath it we find large pockets of relatively unmixed waters, which can flow, for example, from the pole to the equator without intermingling much with surrounding waters.

Once the tritium (or tritiated water) falls into the ocean, it is rapidly distributed throughout the mixed layer. Since this layer is three hundred to six hundred feet deep it contains much more water than is present in the atmosphere above it; therefore the concentration of tritium is much lower in this layer than in the water vapor of the atmosphere. This clue gave Ostlund an idea of how to trace the origin of the hurricane's atmospheric currents: measure the tritium content of the hurricane's winds. Where the water vapor in those winds contains a high amount of tritium, it must be uncontaminated by water evaporating from the ocean. Where it contains a lower amount of tritium, it has been diluted by oceanic water.

It was simple to conceive, but not quite so simple to accomplish. The government provided funding through the National Science Foundation and space on two DC-6s that the Environmental Science Services Administration was flying through the hurricanes. Because of the need to collect many samples on each flight, the cramped space on the airplane, and the rough flight to be expected, they couldn't simply stick a bucket outside and collect liquid water from the rain. Instead Ostlund investigated a number of absorbents and found one that would suck up the water efficiently and keep it in a concentrated and nonspillable state. An external scoop that fed into a sieve trap was mounted on the DC-6; the water scooped in was trapped there, and the sieve traps could be rapidly changed as the plane flew through different regions of the hurricane.

Once back in the laboratory the water was released by heating, and the tritium was changed into gaseous hydrogen form, pumped into Ostlund's special counters, and its radioactivity measured. The result clearly showed that as the plane flew toward the center of the hurricane the concentration of tritium in the water vapor continually decreased, indicating increased contribution from oceanic water; that is, increased dilution of the atmospheric water vapor by evaporation from the ocean. This is precisely what was expected from the general model of hurricane circulation. But then at the very center of the storm the tritium concentration unexpectedly jumped up; this is clearly the signature of an input directly from the atmosphere. Ostlund saw here the sign of water vapor descending from above rather than rising from the oceans below.

5

These experiments, and others like them, were not quickly or easily integrated into our body of hurricane knowledge. People who study storms are meteorologists, those who study oceans are oceanographers, while those who know about radioactivity and its measurements are chemists or physicists. These people do not easily speak to one another. For one thing, different personalities congregate into the different disciplines, although this is a subject I can't discuss in public without losing friends. For another, each discipline has its own jargon and its own paradigms. But on a time scale similar to that of the transfer of tritiated water from the stratosphere to the troposphere—slowly, slowly—the work of one group gets assimilated by the others until today we know pretty well how a hurricane works.

How to stop it is another matter.

6

THE FURIES

The answer, my friend,
Is blowin' in the wind . . .

—from "Blowin' in the Wind,"
by Bob Dylan

1

It would be nice to stop hurricanes at will, or at least to control them so they don't grow too horrendous. Come to think of it, it would be nice to control all sorts of weather: to make it rain during a drought; to stop the rain during a flood; to warm the air when the blizzards roll in across the plains and to cool the air when summer broils us.

The dream is an ancient one. Primitive man's religious hierarchy, in fact, grew largely because the shamans claimed some power over the forces of nature. In at least one case, that of ancient Egypt, this religious power was based not on god but on science.

Egypt has a special relationship to geography and the environment: it is a barren desert except for the waters of the Nile, which pour out mysteriously (at least, it was mysterious in ancient times) from the impenetrable desert wastes far to the south. The people of ancient Egypt had no way of traveling to the source of the Nile, and no way of knowing there were mountains farther south that gathered the equatorial rains and fed them to the Nile. All they knew was that these waters brought them life.

For not only did the river make agriculture possible along its banks, but every year—in the hottest, driest season, when the fear would grow that all the plants would wither and die—the Nile would mysteriously but dependably rise up and overflow its banks, inundating the surrounding soil and making it fertile for another year. If ever this mechanism had failed, the people would have died of famine.

It never did fail, but each year as the season grew hotter and drier the tension must have mounted: would the Nile once again yield to prayers and incantations and rise up to bring life to Egypt? The Egyptian priests took advantage of this most natural fear when they learned that they could predict the time of overflowing by observing the sun and the star Sirius. When the sun and Sirius rose closely together, it signaled that the Nile would overflow within the next few days. And so they observed the rising of Sirius each night, and as daily it appeared closer to dawn they would begin the rising crescendo of ritual and prayers until finally the match was made; then they would announce that the great gods had heard their prayers and the Nile would soon overflow—and so it did, to the great joy and thankfulness of the multitudes.

Of course it would have overflowed its banks with or without the prayers or the intervention of the priests, but who was to know this? And so the people were grateful to the priests and brought them gifts, and they all lived happily ever after.

To this day, even in civilized societies, there remains the hope and belief that prayers and priests can influence the weather, can make the Nile overflow or cause the rains to come, can mitigate the blizzard and mollify the heat. *The Rainmaker* was a popular play and movie of not too many years ago, and reflected practices that still go on: whenever the rains don't come when they should, newspapers report priests and shamans, saints and charlatans—and of course politicians—praying and singing and dancing to bring on the rains. "We can only pray to Allah," the president of Pakistan said when the typhoon hit there, and the president of the United States echoed this sentiment when Andrew hit us a few years later (using a more generic term than "Allah," of course).

All manner of superstition has grown up around the weather. Nearly two thousand years ago, Plutarch told the people that "It is a matter of current observation that extraordinary rains generally fall after great battles," and this belief has persisted to nearly the present time. It gained some scientific credence, or at least pseudoscientific speculation, after the invention and widespread use of gunpowder, when it was suggested that the heavy sound waves of the explosions rocked the clouds and shook loose their stored water.

This belief has a strange life of its own, even though there has never been any evidence for the effect. In the early 1950s a French general of artillery, F. L. Ruby, invented inexpensive rockets for the use of European vineyard and orchard keepers in combating hail. The idea was that they could break up and disperse the hailstorms that threatened their crops by the sound effects of the exploding rockets. Although there never was any proof that the system worked, Ruby became phenomenally successful at selling the rockets: "It seemed that every prosperous grower in France and Italy had rockets to shoot at menacing-looking clouds," writes Horace R. Byers, Distinguished Professor of Meteorology at Texas A&M University. It didn't help—the hailstorms came when they pleased as they always had—but the rockets weren't entirely wasted. As Dr. Byers notes, "The psychological effect of thus being able to shoot at the enemy must have been very satisfying."

During World War Two the unusual weather—as Dr. Byers points out in his excellent review, the weather *always* seems to be "unusual"—was ascribed to gases emitted from munitions factories. Ten years later, when I was a student at the University of Florida, the first rockets were sent into orbit, and I remember reading in the local newspaper that the cold weather of that winter was brought on by these rockets: they blasted holes through the atmosphere, and the "cold air" from outer space poured down through these holes and ruined the crops.

This is not to say that the weather is beyond human understanding and even influence. One need only remember James Espy and his "crack-brained" ideas. He proposed the first reasonably scientific attempt to modify the weather. In the April 5, 1839, issue of the

National Gazette and Literary Registar of Philadelphia he suggested a means of inducing rain when the weather was humid but droughty. He said that if sufficiently large fires were built they would generate updrafts, and in a humid atmosphere these rising currents would be holding water vapor that, upon cooling off in high altitudes, would precipitate as rain.

It was a good idea, but the available technology was insufficient at the time to realize it. It was tested more than a hundred years later, but in the middle of the twentieth century the technology was still lacking, and no encouraging results were obtained. Oh well. Maybe someday.

After the Civil War a man named Edward Powers published a book called *War and Weather* that listed numerous times when rain followed a great battle. Of course you can do this for any war if you: (1) don't set any definite time limit that constitutes how far after the battle the rain must come in order to be listed, and (2) omit the battles that were *not* followed by rain. (This is the method used to buttress faith in astrology: remember the predictions that come true and forget those that do not.) Powers's list was impressive enough to people without any education in probability and statistics, which then included (as it does now) the members of Congress, who in 1890 appropriated ten thousand dollars for experimentation in setting off explosives to bring rain. (Of that amount, nine thousand dollars was actually spent on the experiment. If today we were spending 90 percent of our research monies on doing the proposed work instead of on administrative costs, we would already have solved many of the problems plaguing our society.)

The experiments were duly carried out and pronounced a failure by the scientists involved, but a success by the media of the day—a sharp contrast to the reactions to modern experiments. Evidently in those days people wanted to hear about the infallibility of science, the new religion of the turn of the century, while today we are more sophisticated and know that science is humbug and tomfoolery. At any rate, although the experiment failed, the news of it rang in people's consciousness and led to the honorable profession of rain-maker throughout rural America, a profession practiced to this day.

By the beginning of the 1960s scientists thought they knew enough about hurricanes to consider disarming them. The development of the hydrogen bomb and the grasp of its awesome power led to a certain amount of hubris. The entire world could be destroyed, people were told; could science do all that but not knock out one little hurricane?

As it turned out, yes, it could not. The concept was as simple as it was abhorrent to those few people who were beginning to think of the environmental effects that were being seen to follow most of our new forays into the battle with nature: Barry Goldwater was beginning to talk about nuking the Vietcong; why not nuke the hurricanes?

Well, it was soon learned that neither one of those things was possible. Nuking the Cong would likely bring China into the war; that is, the use of nuclear weapons was seen as a horrendous escalation of normal warfare, which would lead to new dimensions of human suffering. On the other hand, the use of nuclear weapons against a hurricane was the opposite extreme, sort of like attacking an elephant with a Swiss Army knife: there just wasn't anything there that could do any meaningful damage.

A Hiroshima-type bomb gives off an explosion of several kilotons (the equivalent in blasting power of several thousand tons of TNT). The new hydrogen bombs being developed erupted with thousands of times more energy, with megatons (millions of tons)—horrendous stuff, enough blasting power to demolish entire cities at a crack. But the energy budget of a good-sized hurricane is measured in terms of *thousands* of megatons. (If you could harness the energy released by just one good-sized hurricane in *one day*, it would supply electricity for all of the United States for nearly an entire year.) Throw an H-bomb at one of those babies and it won't even sneeze.

So, no, we don't have enough energy at our disposal to go head-to-head with a hurricane, and even if we did it would do so much damage to the environment that we might be worse off afterward than we would be just suffering through the storm.

The next suggestion seemed more reasonable: instead of trying to overpower and destroy the hurricane, why not just steer it somewhere else? After all, most hurricanes never hit shore. They originate over the open ocean, they blow westward over the open ocean, they turn northeast and then back over the open ocean, and finally they reach cooler waters, run out of fuel, blow themselves out, and disappear. Why not just steer the few dangerous ones away from land?

It looked reasonable at first glance, but not at second. The hurricane is not a lumbering monster that can be tipped one way or another like a drunken fullback; it doesn't lumber along at all, at least not under its own power. The direction it takes is dictated by the steering currents, which are wholly (or nearly so) independent of it. Trying to steer it is kind of like trying to steer a cork bobbing in a rushing river. It would be easy to nudge the cork a bit, and so it is, but that nudge won't change its direction. You can push it and shove it as hard as you like, it's still going to run downstream because it has no life of its own, but simply follows the rushing waters of the river. If you want to stop a cork from floating from St. Louis to New Orleans you either have to hope it will brush against the bank and get stuck there or you have to divert the course of the Mississippi. And so it is with the hurricane: to change its direction means changing the steering currents, which are as large compared to the hurricane as the Mississippi is to the cork. (Well, maybe there's a bit of hyperbole here, but you get the point.) It is impossible to know where to start, and equally difficult to know where the whole shebang would end: at a minimum, the weather over the entire hemisphere would probably change unpredictably, and therefore unacceptably.

Obviously, redirecting hurricanes is not an option. It would be wiser to outsmart the storm, which after all is the sort of thing man has been doing to other natural enemies ever since we moved out of the caves. Our teeth weren't as sharp as a saber-toothed tiger's, our claws were laughable—good only for scratching each other's backs—and our muscles weren't as strong; but we're here and the saber-tooth is gone. Man is smarter than anything else in nature, and so we've subdued them all. What we did to the saber-tooth could be done to the hurricane.

It was necessary to find an "Achilles' heel" strategy, as Robert Simpson—who came up with the first plan—put it. He sought to find a technique "whereby a relatively small amount of energy is applied strategically to alter the energy-transformation processes (of the hurricane), the local rates of kinetic-energy production, and the maximum winds that can be generated."

At this point, instead of trying to skip the basic science and attack the hurricane head-on, scientists began to realize the need to start at the beginning, where all knowledge starts, with basic physics and chemistry. That is, to understand all sorts of clouds and rain processes before attempting to bludgeon the most ferocious of them into submission. Luckily, just within the previous couple of decades, we had begun to acquire precisely the kind of knowledge that is required.

The method Simpson ultimately came up with was cloud-seeding, which depends on the concept of supersaturation. You probably remember from a mostly long-forgotten chemistry course what happens when you add to a beaker of water the two components of a salt, such as AgCl, or silver chloride. Adding small amounts of the silver and chloride ions causes them to dissolve, but only up to a point. When the combined concentrations of silver and chloride ions in solution grow large enough, the stuff precipitates out. (Precipitation also can be achieved at lower concentrations by changing the temperature, usually by lowering it.)

Every solute-solvent combination has its own limit to how much of the solute can be dissolved in the solvent, and the dissolution of water vapor in air is no exception. The term "humidity" refers to the measurement of how close we're getting to this limit. One hundred percent humidity means the air is carrying as much water vapor as it can; any increase in concentration or decrease in temperature, and the water will condense out as rain.

What happens in supersaturation is that, under certain conditions, more of the solute can be dissolved in the solvent than theory would predict; that is, the concentrations of the species in solution can rise above the solubility product. A general explanation, in anthropomorphic terms, is that the stuff wants to come out of solution but doesn't know where to begin: no molecule wants to be the first to start something as revolutionary as a phase change. But if at this point the proper seed is dropped into solution, the supersaturated solute will flock around it and precipitate out. It's like dropping Al Sharpton into a milling crowd: suddenly all inhibitions and emotions are released and tumult ensues.

To be more accurate, this discussion about precipitation should be taken one step further, since reality is often a bit more complex than we'd like. In fact, the concepts of solubility, saturation, and supersaturation had been around for a long time, but nobody had applied them to the problems of disturbing nature until one of the foremost American scientists took a mountain-climbing trip.

2

In the early 1940s, when most scientists in America were working on the atom bomb or radar or other wartime inventions, Irving Langmuir took a vacation to climb Mount Washington. Langmuir was one of the few American Nobel laureates at that time (the first, Albert Michelson, had won only thirty-five years before). Although his Nobel work had had nothing to do with the atmosphere and its workings, Langmuir had become interested in the physics of clouds. This might not have led to anything useful had he not also liked mountain climbing and happened to have an assistant, Vincent Schaefer, who was equally keen on the activity.

> "You will see that such things are very important," Langmuir told the National Academy of Sciences a few years later, "and that they are closely related to our subsequent interest in weather control. Many things we are now doing have originated from that common background. Nobody could have planned that for us. If we had not been naturally inclined that

way, no one would have prevailed on us to go up on Mount Washington in the wintertime [and thus begin] the research that occurred as a natural development [of that trip].

"We found, much to our surprise, that anything exposed on the summit of Mount Washington during the winter when clouds were there, as they are most of the time, immediately becomes covered with ice. This was due to the presence of supercooled water droplets."

This was the clue, the beginning of all subsequent work on weather modification. Supercooling of water is a phenomenon closely related to supersaturation. In the latter case the concentration of a solute exceeds its theoretical limit but doesn't precipitate out of solution until either a suitable nucleating seed is added or the concentration is increased until finally it is forced to drop out. Supercooling is much the same thing, but in regard to temperature instead of concentration. Cooling liquid water to zero degrees Centigrade will freeze it under normal conditions; that is what defines the freezing point. But if the cooling is done under unusual conditions — such as keeping the water absolutely still so there is as little molecular movement as possible, or when the water consists of small droplets distributed throughout a cloud — the temperature can be taken well below the freezing point without the water solidifying. Similar to the concentration case, no molecule wants to be the first to go, so they all hang around waiting for another to start the process.

Langmuir and Schaefer discovered that this is a common occurrence in clouds; in fact it is the usual behavior. "The thing that struck us most," Langmuir told the academy when he reported on his research, "is that if there are any snow crystals in the cloud, they will be growing and falling. But in the winter-time, if you see any stratus clouds from which no snow is falling, even though the temperatures in the clouds are below freezing, there just simply are no [ice or snow] crystals there in any reasonable number. Such clouds consist of water droplets. They can be supercooled to very low temperatures."

Most people think clouds consist of water vapor, which upon

cooling turns into water droplets that fall as rain. But this isn't quite so. Water vapor, or steam, is invisible; it doesn't react with visible light, and therefore cannot be seen. (The "steam" seen coming out of a teapot's spout is not steam, but water vapor condensing as the steam cools below 100 degrees Centigrade. That's why it isn't visible at the lip of the spout, but only after it has traveled a few inches and cooled.) Yet we see clouds quite clearly. This is because clouds form when water vapor condenses into water droplets, just as a teapot's steam does. Liquid water is not invisible; it reflects light—think of the sun or moon shining on the ocean—so that the tops of clouds are white. If it reflects enough light the bottoms of the clouds become very dark indeed (since very little light is coming through), and these clouds are loaded with water droplets that will soon turn into rain. The rain process can begin either when the small droplets of water grow large enough for gravity to pull them down or, frequently, when they freeze into snow or ice crystals. The falling crystals may then melt as they reach lower altitudes and reach the ground as rain. Much of our rain actually begins at the cold temperatures of high altitude as snow and ice, even when the weather down low where we live is quite warm.

And many of the clouds we see that never turn into rain actually carry a heavy load of water, supercooled below zero degrees Centigrade, but they don't rain on us because the water droplets don't grow big enough or cold enough to precipitate as either water or ice. Once we understand this, the concept of turning the clouds into rain is obvious.

Actually, the first attempt at rainmaking came before the theoretical concepts were fully understood, in 1921, when Dr. E. Leon Chaffee, a professor of physics at Harvard, sprinkled sand particles over clouds in an effort to provide them with nucleating seeds. It didn't work; in fact, the results suggested that the clouds were actually dispersed, and the media took up the idea that the professor had found a way to provide perfect weather for picnics on the Fourth of July. In actuality, the experiment simply failed: no causal effect was measured.

Langmuir's assistant, Vincent Schaefer, found the modern approach to rainmaking quite by accident, just as the two of them had found the initial idea on their mountain-climbing expedition. Following up on their initial observations, they had purchased one of the first home freezer units and were trying to create artificial clouds in it. Schaefer simply leaned over and breathed into the open unit and saw "ordinary cloud particles" form in it as fog from his breath. But even though the temperature in the unit was 23 degrees below zero, no ice crystals formed. The research eventually hit a standstill, with no further insights, until July 12, 1946.

It was an unusually hot day in Schenectady, New York, where the two were working at the General Electric Research Laboratory, and Schaefer couldn't get the temperature of the freezer low enough for his experiments. So "he took a big piece of dry ice and put it in the chamber to lower the temperature. In an instant, the air was just full of ice crystals, millions of them."

Schaefer jumped into this new result with both feet. He soon found that "even the tiniest piece of dry ice would fill the cloud with crystals. . . . The effects to be seen are wonderful to look at and it is a simple matter to duplicate all the natural conditions of an actual cloud in the sky."

Several months later Schaefer, tired of "duplicating all the natural conditions" in the laboratory, took to the air to visit the natural conditions themselves. He scattered three pounds of dry ice into a cloud at fourteen thousand feet and saw the cloud turn into snow before his eyes. "I turned to Curt [the pilot]," he later wrote, "and I said, 'We did it.'"

Not quite. Further experiments did not always have the same immediate result, and the excitement of that first success dimmed as repeated attempts to turn clouds into snow or rain failed. Evidently the dry ice lowered the temperature enough to precipitate the ice crystals in the small freezer in the laboratory, and by chance in that first cloud, but usually the volume of the cloud was too large to be affected. And then another accident led to further progress.

In the summer of 1946 an MIT researcher, Dr. Bernard Vonnegut, who was interested in the problems of icing on aircraft wings, happened to be spending some time at the GE lab in Schenectady, where he was understandably impressed with Schaefer's work. It turned his attention to the problem of the nucleation of ice crystals. He thought nucleation might be accomplished not by the difficult feat of lowering the temperature of the entire cloud, but by introducing the proper seed crystal instead. The question, of course, was what would constitute the "proper" one? Vonnegut reasoned that a crystal similar to ice in shape and structure was needed. X-ray crystallography is the technique that can decipher the smaller-than-microscopic lattice structure of a crystal, and by the mid-1940s research on this had resulted in a handbook that listed all the known structures. Looking through the book, Vonnegut found that silver iodide provided a very close approximation to an ice crystal; further, it had the necessary characteristic of not being soluble in water (if it had been, it would simply have dissolved in the cloud's water droplets, and effectively disappeared).

Schaefer was away on a trip, so Vonnegut visited the GE chemical storeroom and picked up some silver-iodide crystals, walked down the hall and into Schaefer's lab, opened his freezer, and scattered the silver iodide into it.

Nothing happened. No ice crystals formed. End of story?

Not quite. A few weeks later Schaefer came back, heard Vonnegut's idea, and became rather excited with the concept even though the experiment hadn't worked. The two of them tried some different iodide crystals, substituting various metals for the silver, again with unsatisfactory results. Weeks later Vonnegut suggested that a jolt of electricity would help: he created aerosol crystals by sparking across an electrode gap. Nothing worked.

Then, for some reason, he took a silver coin from his pocket and used that as one of the electrodes. When the spark leapt from the other electrode to the coin, sending a shower of microscopic silver particles into the freezer box, all hell broke loose: the box was suddenly and spectacularly filled with ice crystals.

Vonnegut was stunned. But then the gods grew mischievous again, and attempts to replicate the effect a day or two later proved futile. As much as he sparked the silver coin, nothing happened.

And yet, something *had* happened the first time. What was different now? He couldn't find anything, but all scientists keep painstakingly accurate records of their research, and when he went over his records, he found the vital clue: in the days before his first silver-coin experiment, he and Schaefer had been experimenting with other iodide salts. Perhaps, he thought, there still had been traces of iodide in the freezer air, and the sparked silver coin provided silver-iodide crystals. When he tried to repeat the experiment a few days later, all trace of the iodide had since vanished. But that couldn't be right, he thought, because his first attempts had been with silver iodide and had failed miserably.

And then he had a flash of insight. It would all be explained, he realized, if another accident had occurred: if the silver iodide he had obtained from the chemical stockroom had been impure. It was unlikely, but possible. He tested a sample, and found that it was in fact 50 percent sodium nitrate. Repeating the experiment with pure silver iodide, he found that it worked as well as it had with the silver coin.

The inevitable occurrence of impurities in nature and in laboratory stockrooms is a real nuisance. I spent two years of my life trying to track down a supposed anomaly in iron meteorites, which turned out to be due to contamination by potassium. And one major reason the Nazis weren't able to produce an atomic bomb during World War Two was that one of their critical experiments was ruined when the graphite to slow down their neutrons turned out to be contaminated with a beryllium impurity. (Taking just these two examples, I guess I can't complain about my hard luck in the meteorite experiment; if the gods were trading one experiment for the other, it was a good enough trade.)

So Vonnegut and Schaefer now knew that clouds could be made to give up their water by seeding. Dry ice worked by lowering the temperature far enough below freezing to force the precipitation; silver iodide promised to work even better by providing a nucleating seed. But, as I mentioned above, the first experiments on real clouds gave mostly negative results. Nevertheless, just one year after Langmuir and Schaefer's initial discovery, the General Electric scientists—augmented by the U.S. Army, the Office of Naval Research, and the National Weather Bureau—formed Project Cirrus, which set out to see what dry ice would do to a hurricane. On October 13 of 1947 they found a hurricane 350 miles off the coast of Jacksonville that was headed northeastward out into the Atlantic. They flew out and hit it with eighty pounds of dry ice dispensed along a hundred-mile track, hoping that something would happen.

And something did indeed happen, although not what they wanted: the hurricane shuddered to a halt, turned around, and blew due west. Two days later it hit Georgia and South Carolina.

Everyone was upset with this unexpected and unacceptable turn of events. The Weather Bureau set to work and searched their records, discovering that forty-one years earlier another (unseeded) hurricane had done the same thing—had stopped in its tracks and changed its path. So it wasn't the seeding that had done it, they explained.

But just because one unseeded hurricane had changed direction certainly did not prove that the seeding hadn't been responsible for this hurricane's movements. Irving Langmuir maintained that it *was* responsible; he pointed out that since only one previous hurricane was known to have changed direction, the overwhelming probability was that any given hurricane won't do that, and since this one did do it, the overwhelming probability is that it was the unusual circumstance—the seeding—that had caused it to do so.

The result, naturally enough, was a reluctance to try further experiments on hurricanes. At least not until more had been learned about them, and a better idea of what seeding might do was established.

3

The strategy developed over the next decade and a half depended both on the Langmuir-Schaefer-Vonnegut discoveries and on subsequent insights into the structure, development, and destructive properties of hurricanes. Starting from the last, we know that there are two major components to the damage potential of a storm: storm surge and wind. The strategy chosen was to concentrate on the wind.

Wind damage is due to the energy of the wind, which increases with the square of its speed. So even a small diminution in wind speed would carry a large decrease in energy and therefore in damage. Wind speed, as has been mentioned, is due to a combination of three factors: one, the pressure differential between the eye and the surrounding area, causing the wind to try to rush in toward the eye; two, the rotation of the earth, which causes the wind to rotate around the center instead of flowing right into it; and three, conservation of angular momentum, which means that the wind speed increases as the rotating air masses get closer to the center.

Okay, for starters you can't change the rotation of the earth, so forget that. Next, it would be great to stop the lower pressure in the eye, but no one has yet found a way to do that. Finally, and obviously, you can't repeal the law of conservation of angular momentum — or wait a minute; maybe you can. It sounds ridiculous, but that turned out to be the strategy chosen. Not to actually repeal the law, and not to break it; these things can't be done. But just as with man-made laws, sometimes you can weasel your way around it.

4

The winds come pouring in from all over the tropical ocean, laden with moisture to fuel the hurricane's inner fire. They are sucked into the center and there they rise, release their fuel, and flow away — but they don't flow *all the way* into the center. At the very center, as we found out from the first airplane flights and the tritium measurements, there is a downflow of high-altitude air. The rising currents are pushed out and concentrated in a ring around the center, and as they rise and dispense their watery fuel they feed enormous thunderstorms in what is called the *eye wall;* this is the real center of the hurricane, this is where it all hangs out. The arrows in Figure 6.1 indicate directions of wind flow, and the shaded boxes indicate growing cumulus clouds:

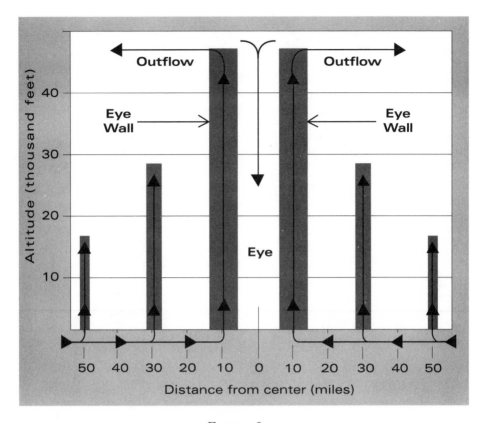

FIGURE 6-1

What Robert Simpson suggested was a Trojan Horse strategy: be like the Greeks, bearing false gifts to overthrow the enemy. Actually *help* the thunderstorm, but in a crafty and deleterious manner, by seeding the clouds just outside the eye wall. Dropping silver-iodide crystals into the second ring of subsidiary storms would induce the water in those clouds to precipitate out, pumping more energy into the system—but not into the *center* of the system. He hoped that the second ring of storm energy would grow at the expense of the inner ring (see Figure 6-2).

This does not change the energy budget of the total hurricane—if anything, it has been increased—but the fire has been fueled at a greater distance from the center. If this is done properly, the original second ring can be transformed into a new eye wall, and the original eye wall will shrink and dissipate. The result is the same central pressure with the same energy sucking in winds from the far reaches of the ocean, but not sucking them in as close to the

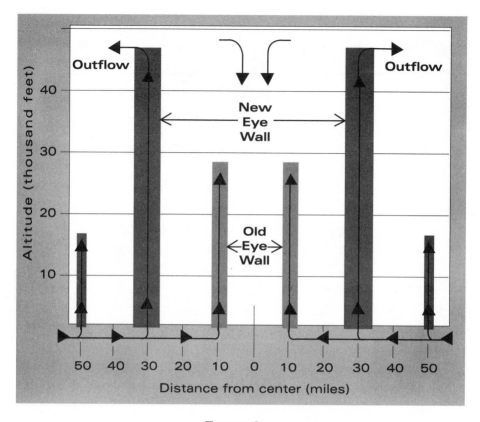

FIGURE 6-2

center as before. The entire storm would now revolve around a larger eye, and if the spinning air masses are farther away from the center, the conservation of angular momentum would not speed them up as forcefully. The wind speed would be less, and so the damage done by the storm would diminish by the square of that amount: if we could cut the wind speed in half, the damage potential would only be one quarter that of the original storm. A category 5 hurricane, the worst possible, would be cut down to one barely deserving hurricane status if its winds could be cut from 150 miles per hour to seventy-five.

Well, that would be, essentially, a total victory, and no one expected such success. But even cutting wind speed by 10 or 20 percent would greatly mollify the storm. So they decided to try. But as soon as they got down to serious calculations, they realized the situation was more complicated than stated above. The maneuver entailed changing the position in which the fuel of the storm, the latent heat of vaporization, was released. This would necessarily change the horizontal temperature gradients, and therefore the horizontal pressure gradients. Three possibilities arose. First, the storm might be unable to cope with this disturbance. Its organization would collapse and it would break up into a bunch of chaotic thunderstorms.

Well, that would be great. But possibility number two turned out to be more likely: it would shrug off the disturbance and re-form the eye wall at its original location. Computer modeling then suggested a surprising result: if the storm did that, the eye would begin to oscillate between the old and new diameters, and this oscillation would add a significant disturbance to the winds, which would pump them up; in effect the hurricane would grow more vicious than it had been before the seeding.

That would be catastrophic, for political as well as moral reasons. In fact, the possibility that seeding could make the storm worse would preclude any action. Suppose a category 3 storm was seeded and these theories were wrong and the seeding had no effect, but from natural causes it subsequently grew into a 4 or a 5? Even though it wasn't the seeding that caused the strengthening to happen, who would believe it? Everyone would blame the scientific "meddling with nature" for creating a vicious storm out of a normal one, just as they had blamed the 1947 attempt for changing its direction.

By the 1960s it was still impossible to tell where a storm was going to go; the 1947 experience was still everyone's nightmare, with amplifications. Suppose we seeded a category 3 hurricane headed for Florida and it turned into a 5 and then swerved around and rammed into Mexico? Every Third World country would protest our "racist" hurricane policy of steering storms away from our own shores and onto those of our defenseless neighbors.

So the most important consideration was to ensure that the storm would not be made worse. It was important to be even more careful than that; like Caesar's wife it was necessary to be above suspicion.

We had to be sure that even if the experiment simply failed to have any effect, at least it couldn't be remotely capable of intensifying the storm. So the scientists involved had to tread very carefully, but the immense loss of life and property caused by hurricanes over the years stimulated them to continue. Finally the calculations showed that if they widened the eye and *kept on seeding it,* the new eye (with a larger diameter) would become dynamically stable. The hurricane would definitely be weakened.

That is, if the calculations were correct.

5

They decided to try. The project was called Stormfury. By September 16, 1961, when Hurricane Esther was blossoming into a category 4 storm located 1,570 miles east-southeast of Miami and heading west-northwest at a steady twelve miles an hour, the team had government funding and a fleet of a dozen Weather Bureau and navy planes ready to go. They flew to San Juan, Puerto Rico, settled overnight, and took off first thing in the morning. They didn't want any publicity, both to prevent raising hopes too soon and to avoid being blamed if something unexpected happened. So the attempt was accompanied by the kind of muddled secrecy that government organizations seem to take delight in. At first, when local reporters noted the fleet of planes landing in San Juan and asked what it was all about, the scientists refused to say. Of course that only fueled the reporters' curiosity, and a few hours later the scientists were forced to concede that they were there to seed the storm.

Meanwhile, a high-pressure zone over the United States was moving eastward. Since it was located north of the storm, if it moved out into the Atlantic before Esther curved northward, it might prevent that curve. In that case Esther would keep moving west and hit land somewhere in the Carolinas. At the last moment, according to the information the team fed to the newspapers, they called off the seeding and simply flew into Esther to take a look around. "They don't want to seed it if there's much chance of it hitting land," *The Miami Herald* quoted the forecaster, "because they don't know what the seeding will produce. If there is any violent reaction they don't want it to affect land areas."

But that was not entirely true. Evidently they simply hadn't made up their minds how to handle the public relations aspect. While they were telling the papers that they had called the experiment off, navy jets were busily dumping millions of silver-iodide crystals into Esther at eighteen thousand feet, as they had decided by this time that silver iodide would be more effective than dry ice. Weather Bureau jets observed from various altitudes to record what was happening.

Then they changed their minds again and decided to admit to the seeding, while at the same time downplaying the attempt. Harry Wexler, chief of research back in Washington, told the *Herald* not to expect "spectacular results. We're not trying to squash the storm, and we don't know how to deflect it. We just want to see how some of its clouds respond to seeding with silver-iodide crystals. We are just feeling our way. The seeding will not decrease the storm's strength," he said, although that was precisely the reason for doing it.

Two days later, on September 18, they had still not decided whether they were seeding it to weaken the storm or indeed if they had seeded it at all. The *Herald* reported in headlines that BEEFED-UP ESTHER CHURNS TOWARD U.S., with her winds now hitting 150 miles per hour, flirting with becoming a category 5 storm, and that "There is no word on whether they [the Stormfury team] have seeded Esther with silver-iodide crystals in an attempt to break her up. . . . Barney Miller of the Miami research center said reports he gets from the expedition 'are only routine . . . of fixes and the like . . .' He did not know why Robert Simpson, leader of the expedition, would neglect to inform him if the storm had been seeded."

Well, the "why" should have been clear enough. It was because the media kept using phrases like "in an attempt to break her up." The Stormfury scientists evidently felt they couldn't get the proper notion of uncertainty and experimentation across to the papers, so they switched to secrecy instead.

By September 19 Esther was threatening the Carolina coast with winds "unaltered by seeding," the papers reported; the team had evidently changed their minds yet again and admitted to the seeding but, as the *Herald* noted, they "carefully avoided the word *failure* in a cursory report issued upon their return to Miami." Robert Simpson "pooh-poohed the report that the team was attempting to change the course or cut down the intensity of the storm: 'It was a small-scale operation . . . a controlled test . . . intended only to determine if we could change supercool liquid to ice. We were apparently successful in our objective. . . .' He said the seeding had no apparent effect on the hurricane," which proceeded to batter the Carolina Outer Banks with twenty-five-foot waves the next day.

Years later Simpson was to write that the experiment had indeed been "aimed at reducing peak winds of the hurricane." He claimed that the eye wall had been temporarily broken open and the wind speeds "diminished," but since they climbed from 130 miles per hour before the seeding to 150 miles per hour afterward, it's not clear what he meant. Any diminution was obviously only temporary and of negligible effect; any victory over the hurricane was in principle only.

Two years later they tried again. Hurricane Beulah came blowing in over the Atlantic in 1963, and on September 13 the Stormfury team went back into action. But they misfired: the silver-iodide flares were dropped in the wrong place, and no effect was seen. Luckily the storm kept blowing long enough for them to get reorganized, and two weeks later they tried again. This time the results were more impressive: the wind speed diminished from eighty-nine miles per hour to eighty miles per hour, as the eye diameter increased from nineteen to thirty-two miles.

The initial funding had been for a three-year period only, but the Beulah results looked so promising that the grant was renewed; and then they settled down to wait.

. . .

They had to pick their hurricanes carefully: first the storm had to be a big one, so that any diminution could be seen, and it had to appear "typical" or "ordinary," without any peculiar characteristics, so that the men studying it felt they had some idea of how it would develop without seeding. Of course no hurricane is completely typical, and the word *ordinary* doesn't even seem to be in the right language for hurricanes, but at least they didn't want anything with obvious peculiarities.

Second, the storm had to develop into a big one far enough away that they could seed it and study it without danger of it crashing into shore. There were two reasons for this, both rather obvious. If the seeding didn't work and the storm grew nasty and then hit an American city, the Stormfury scientists didn't want to sit around and read articles in the newspapers about how meddling with nature only makes things worse, and is this what they're spending our tax dollars on? And, perhaps less obvious, there is the fact that when a hurricane hits land it changes: it's no longer sucking up water, and its motion over land involves different frictional forces between wind and surface than are involved over water. So, one way or another, it changes naturally, and that makes it difficult to see if the seeding has had any effect.

Finally, although the storm had to develop far enough from land, it also had to develop not *too* far from land, because the land-based planes assigned to Stormfury had a limited range. In all, the criteria were rather narrow, and several years passed by after the funding was renewed without a decent hurricane in view. Finally Camille came blasting over the horizon, but she grew and moved too fast, and wreaked her wrath on the mainland before they were ready to try anything. Then suddenly, in her wake, they discovered Debbie.

On September 16, 1969, Debbie was fifteen hundred miles east of Puerto Rico, sprawling over a two-hundred-mile-diameter area and growing winds of "at least fifty miles per hour," according to the Hurricane Center at Miami. The people there reported satellite observations that indicated a "quite well-developed storm [which] is expected to grow into a hurricane rapidly." Moving east at fifteen to twenty miles per hour, it was "expected to move within range of landbased hurricane hunter planes by midday."

"The Red Cross," the *Herald* concluded its August 16 report, "has ordered special disaster teams to possible danger points around the Gulf Coast from Key West to the Texas border."

During the next two days, while Camille gathered public attention by moving into the Gulf and crashing against the shore, Debbie kept the scientists' attention as she moved northward and intensified to full hurricane status with one-hundred-mile-per-hour winds. By Monday morning, August 18, she was five hundred miles east of the northernmost Leeward Islands, the island chain stretching southeastward from Cuba and Puerto Rico. Her motion was northwest, and so she was not in danger of hitting any land within the next few days, although she had come within range of the navy's land-based seeding planes at San Juan. They decided to go for her.

Sixteen planes flew out from Puerto Rico and dropped a half-ton of silver iodide just outside her eye wall. They first reached her at 10:15 A.M. and continued the bombardment all that day and into the night.

The next day the *Herald* reported that immediately after her first seeding, Debbie veered sharply northward and then shuddered to a complete stop, as probably did the hearts of the scientists of Project Stormfury as they remembered what had happened in 1947. But "its intensity hasn't changed," the forecaster reported. "It hasn't shown any sign of either weakening or increasing."

The project scientists were sure that the northward turn and the complete stop had nothing to do with the seeding; they had learned enough about steering currents since 1947 to be certain that nothing they had done could have affected them. But the change of direction was an indication of the problems facing the project, because how could they convince the public that this was true? The project was lucky that Debbie turned northward and stopped instead of suddenly turning west and speeding up. If the Stormfury scientists did anything that could have been interpreted as changing the storm's course toward land, someone would have been sure to shout it from the rooftops.

As it was, the planes seeded all day and interspersed each seeding with observations at various altitudes in an effort to find out if they were having any effect at all on the storm. By the next day Debbie had started moving again, "apparently unaffected by the chemical assault." She turned slightly northwest, increased her speed and increased her winds, but the scientists tried to impress upon the media that this was not their fault: "There are such minor fluctuations in all hurricanes . . . nothing we couldn't relate independently to the natural environment."

Meanwhile, no one knew quite what to expect. Robert Simpson, the Stormfury father and director, kept pointing out that this was an exploratory experiment designed to gather data, not to weaken the hurricane. But at the National Hurricane Center forecaster Gil Clark was telling reporters that it was too early to tell if "Stormfury's objective—to weaken the hurricane—had been achieved." And Harry Hawkins, the project's associate director, was claiming that it was "meteorology's man-on-the moon day" because it "marked the first full-scale experiment to control a hurricane's violence." Clearly the people were bifurcated on the horns of every scientist's dilemma in these days of public funding: generating enough enthusiasm for the project to ensure continued financial support, while at the same time not taking blame if it didn't work.

So the Stormfury people had to tread softly, claiming enough to whet interest but not too much to invite the word *failure.* In the end they did fail, although less in science than in public relations.

Meanwhile Debbie was churning away again. By August 21 her winds were up to 125 miles per hour and she was 850 miles northeast of Miami, heading northwest at a steady fifteen miles an hour, and Stormfury decided to go after her again. All day the navy's A-6 Intruder jets swirled into the maelstrom and released billions of silver-iodide crystals, nearly a ton of them, while the scientists themselves vainly tried to explain what they were doing. Sometimes it seemed as if their message got across: the *Herald* reported that "taming the hurricane is a far more ambitious goal than Project Stormfury set for itself;" the scientists would be merely searching through the data for "some clues about future seeding operations." But that statement came in the closing paragraph of the story; in the lead paragraphs of that same article the public read that "If the ton of silver-iodide crystals dumped into Debbie's eye doesn't weaken the storm, it could mean 'back to the drawing board' for the hurricane-control researchers."

In the end, it was difficult to see any immediate results. But Debbie swung out to sea again and disappeared, while the public forgot about her in the wake of Camille's horror.

The scorecard was muddled. The official progress reports reported, well, progress. The winds in each of the seeded hurricanes had diminished, in the case of Debbie by up to 30 percent. But the newspaper reports gave quite a different impression. The 30 percent reduction with Debbie, for example, was for the seeding of August 18, and according to the official National Hurricane Center reports that appeared in the newspapers, the winds were 115 miles per hour when the planes attacked, and 120 miles per hour the next day.

What seemed to be happening was that the winds were temporarily reduced by the seeding, but within hours picked up again. Apparently Stormfury had messed up the eye wall for only a short time. They were like a virus attacking the human body: at first the body is weakened, but then its antibodies take over and soon it's as healthy as ever. The silver iodide was nothing but a mild cold virus as far as the hurricanes were concerned.

Still, they had done *something*, and they were eager to try again. But nature didn't cooperate. No more suitable storms erupted in 1969, and none at all in 1970. They would run out of money by the end of the 1971 season, and when it opened with a whimper instead of a bang, they were melancholy. As the summer streamed by with nothing in sight they grew more and more worried. When Hurricane Ginger finally appeared in late September they decided to go after her, even though she was not much of a candidate.

She was a loosely organized storm with winds just barely hitting ninety miles per hour when they attacked. They didn't expect much in the way of observable results, and tried to say as much. "NOAA [National Oceanic and Atmospheric Administration] spokesmen stressed that the objective is not to break up or steer the storm," the *Herald* dutifully reported on the first day of the mission, but by the next day it could no longer restrain itself and announced that the project was "a long-planned attempt to bomb a hurricane into submission."

As such it failed, and that was the impression the public received. In his final report Dr. Simpson pointed out that in every hurricane that was seeded the winds diminished, although as I've said above the effect was temporary. In the end no one was convinced, and Project Stormfury's mission was changed to one of observation and analysis only, since the arguments against modification appeared —particularly to the layman—a good deal stronger than those in its favor.

6

The arguments against trying to mess with nature are inherently strong precisely because of that wording: trying to "mess with nature" seems obviously wrong. There is a deep-seated religious conviction in many people that God is running this world and we shouldn't attempt to usurp His position. Never mind that we do that every time we take a whiff of anesthetic or swallow an aspirin, or whenever we irrigate a field or build a dam or wear a condom. Each of these practices encountered resistance when it was first introduced; we should accept the evil we know, the logic somehow goes, rather than fly to others we know not of.

It's not that bad an argument, actually. It has often been found that trying to modify nature on a large scale to suit our own purposes ends up making things worse because we simply don't know enough about how she operates to confound her successfully. Malaria is a perfect example.

Although it has been successfully eradicated from the United States, malaria remains one of the great scourges of the planet. The World Health Organization estimated in the summer of 1990 that about 270 million people were infected with the disease, and each year about 1.5 million children under the age of five die from it. More than 90 percent of the people living in sub-Saharan Africa are infected, and the story of our fight against the disease is a potent illustration of our power and our limitations when confronting nature head to head.

An enormous campaign to wipe out malaria was launched in the 1950s by dousing India with DDT, in an attempt to eradicate the anopheles-mosquito population, which spreads the disease. At first it appeared to work. The number of cases there, which had held steady at about 50 million since accurate counting had begun, dropped to fifty *thousand* by 1961. But instead of continuing to decline to zero, the incidence of the disease leveled off and then began to climb again. By 1965 there were 150,000 cases, by 1969 350,000, and a decade later the number of malaria cases had grown right back to the 50-million level and was still climbing. Today there is more malaria in India than ever before, and with nearly 300 million people infected worldwide, the campaign is remembered as an infamous failure.

The reasons are twofold. First, the DDT killed the anopheles's predators as well as the mosquito itself. Second, the anopheles developed a resistance to the chemical more efficiently than did its predators. So after a few years of spraying, the mosquitoes were no longer being killed efficiently by DDT, and as they began to reproduce again they found that their larvae were swarming into a world free of natural predators. It was as if Someone had said to them, "Go forth, be fruitful, and multiply."

There have been successes to counterbalance this sad story, such as the worldwide eradication of smallpox, so there is room for optimism. The truth, I would like to suggest, is that when we try to change nature we are not doomed to fail but neither are we guaranteed success, and so it is quite proper that we tread softly, go slowly, not attempt to reach beyond our grasp. That doesn't mean we shouldn't try, but when the Stormfury team argued their case before the public, other voices put in a word of their own, pointing out, for one thing, that hurricanes are not all bad. Like the villain in an adult western, they have their good points.

Take water, for example. In addition to the terrible destruction of the storm surge, hurricanes bring rain, and while torrential rains can cause floods, milder rains mean survival to many people. In the long run, and over large areas of the American continent and offshore islands, the rain brought by hurricanes does more good than harm. An early study of Puerto Rico discovered that only 10 percent of the hurricanes that hit the island between the years 1899 to 1928 were "overwhelmingly destructive." Thirty percent brought destruction to local areas but were beneficial to others, and fully 60 percent were overwhelmingly beneficial. The total effect of hurricanes was definitely positive, at least during this period, and there seems little reason to doubt the continuation of this conclusion.

The same holds true of Mexico, where without hurricanes there would not be enough rainfall to support the country's agriculture or to provide sufficient freshwater to its citizens. Indeed, Mexico protested our government's attempt to intervene in the natural process of hurricane formation; it demanded that Project Stormfury be called off. While hurricanes do indeed cause terrible damage, the effect is occasional and local; the good they do lives after them, over the length and breadth of the region and for enduring periods of time.

The conundrum is a delicate one. To steer all hurricanes away from landfall, or to weaken them below their potential for destruction, would deprive our continent and islands of rainfall, which, as I've said is necessary for agriculture and for freshwater supplies. But every few years a hurricane strikes with such devastating force that the obligation to try to protect people living under the shadow of potential disaster is pressing. Project Stormfury tried to assist in the right way: They seeded hurricanes far from land, trying to induce changes that would be so small they could be recognized only by careful statistical analysis after the event. Unfortunately they ran into criticism from both sides: the media played up the project as an attempt to "bomb the hurricane into submission," and, when it failed, the public and Congress turned against this waste of money. The project was accused both of attempting too much and accomplishing too little. In truth, they attempted very little and, if they did not actually accomplish it, they showed at least promising results.

That hurricane winds diminished and the eye wall changed temporarily was a good beginning to a worthwhile experiment, but the project failed on political grounds. To continue at the same conservative pace, preventing neither the disastrous hurricanes nor the normal distribution of hurricanes, would mean they could learn more and more about how to control them. Eventually enough would be known to permit an attack on only specific hurricanes that pose a definite threat to our coastal shores.

Alas, the Stormfury results were too subtle for this to be recognized and accepted, and the hurricane seeding experiments have been put on hold to this day. Still, although people can be stopped from spending millions of dollars on expensive experiments, they cannot be stopped from thinking, and some interesting alternatives to control a hurricane's fury have been devised.

Dr. Simpson and his wife, Joanne (nee Dr. J. S. Malkus), have suggested attacking the hurricane's source of fuel. The energy that powers the hurricane is derived by evaporation of warm water from the ocean surface. So the Simpsons came up with a plan to cut it off by coating the ocean ahead of the advancing storm with a layer of nonevaporative film. The effect would be like spreading a plastic carpet over the water so that when the hurricane hit it, the fuel source would be cut off. No water could be sucked up into the vortex, the air rising there would remain dry, no latent heat could be provided to the storm since there would be no water vapor to condense, and the storm would peter out just as a car engine dies when it runs out of fuel.

It was an interesting idea. There were just three drawbacks. One was the cost: the Simpsons calculated that they would need a fleet of about thirty giant planes (C-130 or a similar type) to spread the film over the water. This was an expensive consideration, but still it would be cheap compared with the cost of a hurricane's hitting a major population center. It was the second problem that knocked out the idea.

Gote Ostlund's tritium experiments are again important. By measuring the tritium content of the air within the hurricane, he could trace the origin of the water vapor that fueled the storm. He could construct a water/energy budget, a sort of water-in/water-out kind of balance sheet. When he did he found that less than 15 percent of the energy of the storm was derived from the water the hurricane actually passed over; the other 85 percent came from waters hundreds of miles away. This gives a quantitative estimate to the rather hand-waving statement I made earlier: that the hurricane is a huge heat pump gathering in the sun's energy from vast stretches of warmed water and pumping it in concentrated form into the heart of the beast.

What it means in terms of slowing down the beast is that even if it were possible to coat the water ahead of the hurricane so that it doesn't get sucked into the maw, even if that was accomplished with, say, 70 percent efficiency, the hurricane's energy would be cut down by only 10 percent. And since the wind speed is determined by the square root of the energy, the wind would be cut by only 2 or 3 percent. Clearly, this is not enough to make much difference.

And finally, the third consideration: the winds of the approaching hurricane would whip up the waters and dispel the film, blowing it away and rendering it useless. Unless someone can come up with a compound that will stay on the ocean surface despite the hurricane winds and ocean turbulence, and unless someone is willing to spread the stuff over many hundreds of square miles of ocean, the idea simply will not work.

Next, Joanne Simpson came up with a proposal approaching science fiction: towing icebergs from Antarctica and putting them in front of a hurricane. The idea is that a hurricane needs warm water to evaporate; unless the ocean temperatures are higher than 26 degrees Centigrade, hurricanes can't develop. An iceberg, sitting in warm tropical waters, will naturally melt. Calculations indicate that it will melt mostly along its submerged vertical walls, and the melted water, although cooler than the surrounding ocean, will rise to the surface because it is salt-free and therefore less dense. The effect will be to cool the surface below the necessary boundary for hurricane formation.

There are some indications that the plan might work, the strongest of these coming from hurricanes themselves. Since they obtain energy by sucking up warm water from the ocean, it follows that the oceans must cool down as they pass over. In addition, they roil up the ocean waters, bringing to the surface the cooler deep waters, and they rain on the ocean, further cooling it. These cooling effects, which have been documented, amount to several degrees and last for ten to thirty days.

So have we seen any effect on subsequent hurricanes? The answer is definitely yes, on the few occasions when a second hurricane followed closely enough in the tracks of the first. In 1955 Connie churned up the Atlantic and cooled the surface waters by nearly 3 degrees. When Diane developed a few days later she crossed Connie's track, and as soon as she hit the cooler water she withered and fell below hurricane force. And when Anita crossed the Gulf in 1977 she cooled the waters by 4 degrees, so that when Babe tried to follow she died like the Wicked Witch of the West under Dorothy's bucket of cold water.

There was another interesting point about Diane. We've already said that although hurricanes cause a lot of damage, the rains they bring are essential. Wouldn't it be lovely if the wind speed could be diminished and thereby the damage, while we retained the water supplied by these storms? Well, that's exactly what happened with Diane. Although her winds died down below hurricane force, she didn't lose her water, and upon arrival in the northeastern United

States (as a storm, although no longer a hurricane) she dumped enough water to cause severe local flooding. That was too much of a good thing, but you can't expect perfection. The point is that the problem of water loss might not be involved in this method of hurricane modification, as it is in seeding.

Of course, one case does not a statistical study make, and there are several other problems involved. The plan would cost a lot. The largest seagoing tugs would barely be able to move icebergs of suitable size, so Joanne Simpson suggested constructing "supertugs," which would cost about $15 million a year to operate. That cost is in 1975 dollars; for comparison, also in 1975 dollars, hurricane damage to the United States approaches $800 million a year in economic costs alone, neglecting human injuries and death and misery. So although expensive, clearly it would be worth it—if it could be done.

Ay, there's the rub. For in that ocean voyage what harm may come is hard to estimate, since the iceberg would be melting the entire way. Ambient water-surface temperatures affect the air above them and thus the climate, both locally and worldwide; inducing a cooling effect strong enough to stop a hurricane in its tracks might also affect weather along the entire track of the iceberg, to an unknown and therefore unacceptable extent.

Subsequent study on the use of icebergs for a variety of purposes has tended to emphasize this problem. It has been suggested that they be utilized to bring freshwater to Saudi Arabia or California, for example, but the anticipated environmental consequences have so far doomed any such attempt. Even if this were not a sticking point, there are practical difficulties in preventing hurricanes with them. For example, how far in advance do you know a hurricane is coming, particularly one that looks dangerous and needs to be stopped? Now compare this advance-warning time with the time required to tow an iceberg many thousands of miles. Remember, most hurricanes are beneficial and should not be tampered with.

No matter how you do this calculation, it doesn't look good. So instead of racing north with a giant iceberg in tow when the radar turns up a storm, how about simply bringing a brace of bergs up for every hurricane season and anchoring them offshore of the big population centers along the Gulf and Atlantic coasts? Then they'll be there when the storms come, and they'll affect only hurricanes that would definitely be hitting land destructively. Although the cost would be high it would be nowhere near the average economic costs presently inflicted by hurricanes. Yet there is still an environmental problem: the effect such bergs would have on local and even global weather, on commercial fishing, and indeed on every aspect of the environment connected with the oceans and the temperature of the air above them. Even Joanne Simpson admitted that the method "will not be practical in the near future" for Atlantic hurricanes, but suggested it might be worth considering for Australia.

As we tend to pay little attention to Australia in this country, almost nobody knows about hurricane Tracy, which "virtually wiped out" the town of Darwin in 1974. Northwestern Australia in general —and its offshore oil drilling industry in particular—is susceptible to the frequent "willy-willies" that strike the coast. And Australia's proximity to Antarctica makes a natural first choice if icebergs are to be used. But that is a big "if," and so far no further work has been done.

In fact, not much work of any kind has been done on climate modification since the first pioneering studies, and absolutely nothing has been done on hurricanes. Cloud seeding has been sporadically attempted in connection with bringing rainwater to drought-ridden regions; the results have been good in terms of water-per-dollar cost effectiveness, but less promising in terms of bringing enough water to relieve drought conditions. No cloud-seeding experiments have been conducted on hurricanes since 1969, and no nonevaporative-film or iceberg experiments have ever been conducted. In recent years science has become dependent on government funding, and scientists have adapted to this new constraint. Most of us have developed a sense of what is currently fashionable, because that is what pays off in funded grants.

The present rate of fundability is terribly low, with less than 50 percent of even the good grants getting funding. In order to have a decent chance at government money a proposal has to be "exciting" as well as good science, and it turns out that the word *exciting* changes its applicability with the volatility of women's fashions. What is "in" one year is "out" the next, and since the up-and-coming young scientist has to present a record of steady funding if he wants to get tenure at a decent university, he learns to feel out the fashions. For the past couple of decades hurricane modification has not been fashionable, although there is potential to protect ourselves from the most vicious and devastating attacks nature can hurl at us. As Robert Simpson wrote, rather bitterly, in closing out his report of the Stormfury project: "Politicians and their constituencies are unlikely to agree on the priority for multiyear funding for such an experiment, until a major hurricane brings virtually unparalleled disaster to a coastal area of the United States."

Which brings us back to Andrew.

7

HURRICANE ANDREW

It was just one of those things,
Just one of those crazy flings . . .

—from "Just One of Those Things,"
by Cole Porter

1

From *USA Today,* Monday, August 24, 1992:

ANDREW HITS FLORIDA

BIGGEST STORM IN DECADES

HOMES IN ON THE COAST

Hurricane Andrew chased as many as a million people from their homes, then descended on south Florida today.

Andrew's edge hit the coast at 1:15 A.M. EST today, threatening to wreak havoc with winds between 135 and 150 mph, heavy rain and 18-foot waves.

Evacuations were ordered Sunday after Andrew hit the Bahamas, killing at least four.

The last bus left Miami Beach at midnight. . . .

2

I am not a happy camper as night falls on Sunday evening. I go outside as the sun drops below the horizon, and I stand there looking to the east. Nothing is to be seen: no lightning in the distance, not a hint of wind. The usual rumbling thunder of wheels on the distant highway is absent. There is no sound in the air but an intermittent hammering. I go back inside and close the door.

It seems so peaceful that it is impossible to believe there is a mon-ster out there coming for us. I turn on the television. "It's absolutely for sure," a very earnest announcer is saying, looking directly into the camera, looking directly at me. "No question about it. It's going to happen tonight." I turn the damned thing off.

We have enough supplies to get through the night and a couple of days without electricity or water. The house is solid enough, I think; it was built in 1966, and through the years there has been no evidence that the builders might have watered the concrete or skimped on the construction codes, but really there is no way of knowing. Some of my neighbors have installed hurricane shutters, others are now trying to nail or screw plywood boards over their sliding glass doors and picture windows. I look at my bare windows with growing regret that I didn't spend the day preparing, running around to hardware stores and trying to find a piece of plywood. Too late now. Hurricane shutters, I thought, were too expensive if you bought the best, and cheaper ones were worse than useless: if they couldn't withstand the wind they would be ripped off to become projectiles or battering rams against the windows they were supposed to protect. Trying to drill holes into a concrete house to put up plywood boards was too hard; I thought I wouldn't do the job properly and would be worse off than with nothing should a board end loosen and begin flapping against the window. I stand now looking at the sliding glass doors leading out to the pool and patio. That is the weak point, I think; if the change in barometric pressure pops them out or if a tree limb knocks them in, the wind will come rushing into the house like a locomotive. And then I re-member the turbine fans up on the roof.

The roof was redone a few years ago, and at the time I installed two aluminum turbine fans to draw hot air out of the semi-attic. They are guaranteed to withstand sixty-mile-per-hour winds; in a hurricane they are supposed to be taken down and the hole that is left in the roof battened over with a metal cover. I go up on the roof but find the attaching screws have rusted shut. This is something Sears didn't mention when I bought the damned things. Nothing I can do will get them loose, and in the end I leave them up there, spinning even now in a terribly weak breeze.

Aside from the fans and sliding glass doors, I think, we are in good shape. *(Yes, but aside from that, Mrs. Lincoln, how did you enjoy the show?)*

The story of the three little pigs pops into my head: *I'll huff and I'll puff and I'll blow your house down!* For the first time I understand what it means and who the wolf really is. I understand now how the two dumb little pigs felt when the wolf came to the door, and I wish I had a brother who had built a good solid house to run to.

3

At 9 P.M. Channel 7 is showing a reporter on Miami Beach talking to the residents. The streets are empty of traffic, and the wind is skittering papers and garbage-can lids down them; in open doorways people are standing around, watching to see what is going on. The reporter says, "I ask people why they aren't evacuating, and they laugh and keep telling me about every other hurricane they've lived through. Well, this one ain't gonna be like any other hurricane you've ever lived through."

4

The television reports are not comforting. That damned storm is not swerving north, it keeps coming right at us. It's now so close that even if it does turn northward we'll still catch a hell of a hit. And they say it's coming faster than they thought. Instead of hitting early tomorrow morning it will come before daybreak. The peak winds should hit around 5 A.M. or 6 A.M. And they're talking now of the winds climbing up to 150 miles per hour, which puts it on the border of a category 5 hurricane.

This can't be right. The first hurricane to hit here in twenty-five years can't be the worst one of all time. That's too much of a coincidence. We're entitled to a small one first, aren't we? Just to learn what to do?

As night falls I go around the house taping shut the windows. They are jalousies and don't close tightly; a strong wind could get in under the tips and lift them up, so I cover the edges with silver duct tape. I bring in all the outside furniture and pile it in the living room, together with the potted plants and the gas grill. The grill will be useful for cooking if the electricity goes; that is, if there's enough gas to last a few days. I try to remember the last time I had the butane tank filled, but can't.

Finally, when everything is inside, I tape shut the front door so the wind won't blow rain in around the edges. Halfway through I run out of duct tape, so I finish the job with Scotch tape. Then I smile at my wife, Leila: "Snug as a bug in a rug."

5

The news on television gets worse and worse. Andrew keeps roaring toward us, and it looks like it really might be a category 5. By ten o'clock the wind is rustling through the streets here, ten miles inland. By eleven it's blowing more strongly than anything I've ever seen. Palm tree fronds and newspapers are blowing along the street, and the rain—although not heavy—is slanting down at an oblique angle. The eleven o'clock news is showing the hurricane just offshore, and the announcer is telling us that it is still coming straight at us, with no time left for it to swerve away. Dr. Bob Sheets, head of the National Hurricane Center, right across from the University of Miami campus, is on the tube intermittently. The Channel 4 announcer now asks him, "You want to pin down just where the eye is going to come ashore?"

"Well, it looks more and more now like south Dade County," he says. "The center will be very close in to the Cutler Ridge area, maybe the Homestead area, but coming right across south Dade County here."

I can't believe this. Who's in charge here? That report puts the center of the hurricane moving across the state just about ten or fifteen miles south of us. And the strongest winds in a hurricane moving due west will be directly north of its eye, where the rotating wind speed and the linear wind speeds combine most forcefully. That puts us right in the path of the most destruction.

I listen to the turbine fans spinning wildly on the roof just over my head. Again, I look at the sliding glass doors, which can be lifted out of their grooves with the slightest of pressure. When the eye brushes past us the outside air pressure will dip way down. Will the inside pressure be enough to pop those doors right out of their grooves and fling them out into the wild wind? I look at my front door, fastened around the edges with Scotch tape. . . .

6

We try to sleep, concentrating on the fact that our house is solid concrete, trying to forget about the weaknesses, hoping that when we wake in the morning it will be all over. Incredibly, within minutes I hear my wife's breathing transformed into the deep, slow rhythm of sleep. I had expected to hear my father wandering around the house all night, testing the windows and the door, but there is no sound from his room or from the rest of the house. Without further thought, consciousness slips away from me.

I wake with a startled shock when a sharp bang rattles the house.

"What was that?" Leila asks, clinging to my arm, upright in bed as I am in that moment.

"I don't know." I listen, but there is no repetition. "Something hit the house." The wind is howling with a frightening fury; it sounds alive, vicious, hungry. I get out of bed and walk through the house. Something must have blown off another house and slammed into us. A roof turbine, a door, a roofing tile? Who knows? If it had hit the window it would have shattered it. The clock says it is just after midnight. There will be more flying missiles; they can't all miss the windows.

We are in trouble. I had never imagined a wind like this one. It's hitting us from the east-northeast, as it should from a storm coming at us from the west. If it were far away to the south the winds would be coming from the north-northeast. This close to easterly is a bad sign: the eye must be just south of us. I turn on the television.

It is no longer possible to pretend the hurricane might swerve north and miss us: it is already upon us. The radar picture shows the eye coming ashore in south Dade County, moving directly west like a murderous Cyclops. It hasn't changed direction one iota since it was first spotted over a week ago. The son of a bitch is inexorable. It is hitting Homestead head-on, and the peak winds should be between there and our house. It's moving even faster than anticipated, meaning ground winds will be stronger. I see a radar picture that looks like the whole universe is swirling around and falling down into Dade County.

The screen goes blank, and a moment later the announcer is back on. He tells us they lost the picture momentarily; he talks to his compatriots and listens through the little headphone set on his head, then he turns back to us. He has just been informed that there will be no more radar pictures of the storm. The National Hurricane Center's weather radar has just been blown off the building and has crashed eight stories onto South Dixie Highway.

I drive past that building every day; it's just across South Dixie from the University of Miami campus. It's a solid quadrangle of stone, built for the Hurricane Center, symbolic of our strength as nature's adversaries. The radar is a huge sphere on top, built to withstand the worst possible storm.

Or so we thought. It's gone now. There will be no more estimates of wind speed, the announcer says: all the National Hurricane Center's wind-measuring instruments have also been blown away. This storm is off scale, the biggest we've ever encountered—

With a click the screen goes entirely blank; its colors fade away into a rapidly shrinking rectangle of gray that disappears before my eyes, leaving me in total darkness. "What happened?" Leila asks, and I realize she has been standing beside me, although I can't see her now. The lights are all off. "Electricity's off," I say.

"Oh, God."

I get the flashlights I had put on the kitchen table and hand her one. She goes to check on my father. I go to the back side of the house, facing west away from the wind and the driving rain, and open the window a crack so the pressure inside the house will match that outside. Hopefully this will help keep the sliding glass doors from blowing out, although I don't know how quickly the internal pressure will equalize with the lower pressure outside as the eye approaches. It depends on how quickly the barometer falls. I move to the east side and check the windows; they seem fine so far, but the carpet beneath them is wet.

I bring back towels to soak up the water, but there's too much. "Dad's all right," my wife says. "He's asleep. What's wrong?" I show her, and we pile towels around the windows.

"Oh, fuck," I say.

"What?"

"The water." With the electricity gone the city water pumps will shut down. I forgot to fill the tub with water so we could at least flush the toilets. I was so happy to find toilet paper at the store, and then I forgot the water! I rush now into the bathroom and turn on the taps. Luckily the water is still running. I turn the bathtub taps on full, put the plug in the drain, and go back to help Leila pile towels around all the east-facing windows.

"What about the other side?" she asks.

"We'll have to move the towels when the wind changes." I know she still doesn't believe me about that. The wind is from the east because that's where the storm is, she insists, and when the storm passes it will be gone. But it won't. The trailing winds will be from the southwest, and although they won't be as strong as the easterlies, they'll probably be bad enough.

Of course it might not matter, we might be obliterated by then. The windows are rattling in their frames as if they might pop out at any moment, even the roof is creaking frighteningly. Luckily no other missiles have hit us, although the wind itself seems strong enough to rip the house open. I shine my flashlight on my watch. It's nearly one o'clock now; hours yet to go. . . .

7

My father is sleeping through it all. Leila keeps checking on him; like me, she can't believe he is sleeping through all this. The sound of the wind is like an endless train roaring by inches from our house. She's afraid he is passed out rather than asleep, but he seems all right.

I keep pacing around the house, checking the windows, waiting for the damn wind to shift so I can close the west windows and open the east. But the wind keeps blowing in the same direction; the only change is in its intensity, which keeps getting incredibly stronger every minute. Like a whistling steam kettle on high heat, the storm winds its whining screech higher and higher until it has to burst.

And suddenly there is a new sound, screaming through the house like a headless ghost. The turbines on the roof, having been ripped out of shape by the wind, are howling in protest, metal rasping on metal. I think, can it be long before they give up and fly away into the maelstrom? I climb up into the crawl space and make my way along the pipes and insulation until I can see them. Miraculously, they're still in place, but sound as if they'll be pulled apart any moment. When they go, a one-foot-diameter hole will be left beneath each of them. Leila passes up a couple of buckets and I fasten them under each hole. They won't be much good because the rain won't be coming in vertically, but there's little else I can do. Up here I can feel the roof sighing, lifting up with each passing gust of wind, like a filament in a pump. What I had never considered before now seems possible: the roof might be ripped loose.

I crawl back down again and try to forget about the roof and the turbines. I stalk through the house checking the windows and the towels beneath them. As I pass by the bathroom door I splash into water nearly ankle deep. I whirl the flashlight around, trying to find where it's coming from. This is an interior hallway; I am standing beside the bathroom door nowhere near a window. I take quick steps in each direction and the water diminishes. It must be coming from the bathroom itself—

The bathtub! I forgot to turn off the water. I clump in and the water is up *past* my ankles. I slosh over to the tub and turn off the faucets, then stand there not knowing what to do, feeling not only scared but wet and stupid, too.

Leila sticks her head around the corner. "What happened?" she whispers.

I start to say something about the bathtub but she shakes her head impatiently. *"Listen."*

I listen; there is no sound but the howling of the wind. "I don't hear anything."

"That's what I *mean*. What happened?"

The turbines have both stopped. They must have ripped loose. I climb back up into the attic and shine my flashlight down its length, but don't see any water streaming in. I crawl up under the vents and shine the light. The turbines are still there. They have stopped spinning; the metal fins are jammed and locked against the casing, which bulges out into the wind, but they are still there. Nothing to be done, I sigh, they'll go when they go.

I climb back down and find my wife standing in the mushy carpet outside the bathroom door. I explain about the bathtub. She gives me that old, familiar look, but says nothing.

8

Three A.M.

Leila seems tired; I'm too scared to be tired. The wind has continued to increase steadily and I wish I had installed good, solid hurricane shutters or at least plywood to board up the windows, but it's too late for that now. I've been trying to change the towels by the east windows, where the rain is coming in, but Leila pulls me away. She's afraid I'll be standing in front of the windows when a tree limb or a garbage-can lid or a roof tile comes crashing through. She's right. I've seen pictures of two-by-fours thrust right through palm trees by storms less powerful than this one.

There's nothing else we can do, so we go back to bed. My father hasn't yet wakened. He's in a south-facing bedroom, and ours is southwest, so we all are as safe as we would be anyplace else in the house. Other than worry, there is nothing we can do, so we might as well get some rest. Leila falls asleep right away, but I lie there listening to the wind still growing stronger. We can't take much more of this. I hear the east windows banging in their metal grooves; the wind alone will blow them loose if it gets any worse — even if no outside missiles come bursting into them — and the wind will certainly get worse. We have three more hours of this, with the wind increasing minute by minute . . .

At 4 A.M. it reaches a frenzy. The whole house is shaking, and so am I. Leila and my father are still asleep. I am making plans, thinking what to do when the windows or glass doors blow out. The house next door has shutters on all its windows, but I don't think we could make it across the lawn. We should have gone there earlier in the evening. We have no interior space without windows, except our bedroom closet. If I took out all the clothes, we could huddle in there. Or in the bedroom hallway, I think; but, no, it's sheltered from the bedrooms only by interior doors, which obviously were not built to withstand any stress. If the wind comes in, these doors will blow out.

But then so will the closet door. No, it opens inward. If we could brace it from within . . . with what? I could empty out our bedroom dresser and drag it inside, then load the drawers again with the heaviest things I can find, and shove that against the door. But with the dresser in there, there will barely be room for all of us. To spend hours in that cramped space with the wind tearing at us would be impossible. It would be like huddling in a sinking submarine, claustrophobic; better to surface and fire our torpedoes at the Nazi destroyers and at least take some of them with us —

I must have fallen asleep after all. Dreaming. I open my eyes and listen. The wind is still beating from the east, but it is no worse than before. I listen carefully, pinning down the sound. It comes directly from the east, no longer northeast. The north-side window is quiet. That means the eye is directly south of us now. I check my watch.

It's still running; somehow that surprises me. How can anything be normal in this maelstrom?

It's only four o'clock. The eye shouldn't be passing us for another two hours or so. I lie in bed and listen. Minute by minute goes by and the wind is terrible, but it is not getting worse. It should be increasing in strength, but it's not.

I click on the transistor radio. By the time I'd arrived at the store yesterday there were no batteries on the shelves, and the one battery in the radio is left over from last year's football season, so I want to conserve it; in the days to come it could be our only link to the outside world. But I listen now and, hallelujah! The storm is moving much faster than anyone anticipated. Instead of five to ten miles an hour it is whipping through Miami at more than twenty miles an hour. The eye is already through Homestead and is heading west across the state toward the Gulf. The increased speed has made the winds stronger than expected, but the hurricane is lasting only a fraction of the time it should have, so the worst is nearly over.

And our roof is still attached and the windows are still in place. I turn off the radio and listen to the rain battering the glass, and soon it seems to be lessening slightly. I'm not sure, I don't trust my ears, but within another half-hour I hear the rain slashing at the southeast windows. I get out of bed to see for sure, and it's true. No rain is hitting the north windows—it's still scudding hard against the east, but it's definitely hitting the southeast windows too. Clearly the storm is passing, and because of its high ground speed Leila was right: the trailing winds are much slower than the leading winds were. For a storm with 150-mile-per-hour winds and a normal ground speed of five to ten miles per hour, the difference between leading and trailing winds normally wouldn't be great. But with a 20-mile-per-hour ground speed the leading winds would be close to 170 miles per hour and the trailing winds just 130.

This sounds good in theory but I can testify it is absolutely, fantastically *great* in practice. Now, lying in bed, I can definitely hear the winds dying down; the rain splatters gently, no longer sounding like the rattle of 30-caliber machine-gun bullets, increasingly hitting the south instead of the east windows.

By five o'clock it is over. It's a bad rainstorm now, but nothing more than that. I don't see any water dripping in through the ceiling, so the roof must have held; even the turbines didn't let in much water. The windows and the sliding glass doors are still in place, and the howling of the wind now sounds only dismal and verbose, like a frustrated marauder sent on his way, disappointed, failed. It's a comforting sound, and I fall asleep listening as it dies down minute by minute.

9

We wake a couple of hours later, with the first thin streams of a hesitant light. We dress and go outside. The rain has stopped, the wind is gone. The sky is gray, unusual for Miami; the air is humid and sticky, normal for Miami. The world is quiet, very quiet, eerie and unreal.

Up and down the street, people are coming out of their homes. They look as if they are stepping through looking glasses into a time warp. Europe after the Second World War.

Debris is everywhere. Broken glass litters the street; it crunches underfoot. We take a few steps down the walk, then turn to look at the house. That first loud bang heard last night is now visible: there is the firm imprint of a roof turbine on the wall, next to the plate-glass window. It blew off the house across the street and smashed into our east wall with enough force to leave its image there forever. If it had hit a foot to the right it would have gone through the window, and Andrew would have followed into the house like Sherman into Georgia.

The skyline is different, and for a moment I can't determine the difference. Then I realize, slipping into another time warp: it looks like it did when the village was first built twenty-five years ago, when the developer erected houses after having razed the natural foliage. There were no trees then, and there are none now. The only things silhouetted against the gray sky are houses. It looks like the neighborhood was redone overnight by Edward Hopper: stark architectural lines, no softening foliage. The trees are all lying on the ground, or stretched into houses with crumbling roofs over them. The house to my right has lost its roof. My neighbor to the left had a magnificent avocado tree in his front yard; it is now sprawled in my driveway, its tips lying a few feet short of my home. If it had been just a bit taller it would have crashed through my roof, just as many trees along the street did to the homes standing there.

Slowly Leila, my father, and I walk along, staring incomprehendingly at the damage, picking our way over fallen trees, avoiding the broken glass and roofing nails when we can, nodding dumbly to our neighbors. It is deathly quiet: no traffic, for a car couldn't drive more than a dozen feet without being stopped by a fallen tree. No one is talking; everyone is staring, turning around, walking quietly on, trying to take it in, trying to understand what has happened.

8

DISASTERS AND WARNINGS

But where are the clowns?
There ought to be clowns . . .

—from "Send in the Clowns,"
by Stephen Sondheim

1

What has happened is the greatest disaster ever to hit the United States. Not in terms of lives lost, for the warning system worked and resulted in the largest mass evacuation in our history; but in terms of material damage and sheer destruction, this was the worst thing that has ever happened to us. I had always thought that a person had to be crazy to live in California, because of the ever-present threat of earthquakes. It seemed unthinkable that people would live in a region where disaster can burst upon them at any moment. I didn't consider hurricanes as equally threatening, but rather as being somewhere between a nuisance and an adventure. I have always loved a good thunderstorm, with lightning crackling and thunder roaring and rain thumping down in buckets, and with me sitting under a solid roof in front of a blazing fire, looking out at the harmless fury of nature. I thought of hurricanes as a sort of super-thunderstorm, hanging around a bit longer, blowing a bit harder. Old-timers in Miami talk about their hurricane parties, when they would gather in someone's house to ride out the storm with a couple of days of drinking and partying. Afterward, the stories went, the streets might be flooded for a day or two, electricity might not be restored for a day or two, and the supermarkets might be empty, but again, for only a day or two.

I now know the reality. It would be two weeks before electricity was restored to my house. During that time we had a trickle of water in the pipes, enough to flush the toilets and even to wash (my bathtub flood was unnecessary). But the lowered water pressure led to a threat of backwash from polluted soil around Miami, so the water was not safe to drink. And without electricity we couldn't boil it. We ran out of gas for our grill the second day, and my attempt to get it refilled found the storehouse with plenty of gas but without power to pump it from the tank.

The screens were torn from our patio, of course, so we had to keep the sliding glass doors closed because of the insects. With them closed there was no cross ventilation, and the usual August and September weather in Miami made life unbearable: temperatures were in the nineties during the day, the high eighties at night, with humidity hovering around 100 percent. Without trees, there was no shade; without electricity, neither air-conditioning nor fans.

The food spoiled in the refrigerator and freezer after a few days, and there was no ice or food available anywhere in the city. We drank warm beer and diet Coke, ate dry cereal without milk, sweated, and waited.

And the worst part was that we couldn't complain: we were the lucky ones. When the *Herald* was delivered again two days after the storm, we found that homes just a mile or two away had been completely destroyed, windows blown out, roofs ripped off, furnishings sucked away with the wind. One family nearby had been huddling in a closet for hours, ever since the roof went; then suddenly the door blew out and the woman's child, whom she had been cradling, was torn out of her arms and blown away. She ran hysterically after the child—out into her broken home, through the hole left where the door had been, out into the storm. Her husband chased after her, carrying a flashlight, and miraculously they found the child alive and bawling, flung into a bush on their neighbor's land. Somehow they fought their way back against the wind to the battered comfort of their home, and made it through till morning.

When morning finally came, thousands of people crawled out

from the rubble that had been their homes and staggered, dumb-founded, into the cluttered streets. One or two of them stepped into puddles electrified by fallen high-tension wires and were immediately killed. Many of them, more than a quarter of a million people, found they had been moved from the ranks of the middle-class into the hordes of the homeless. Frank Morkill and his wife, Nancy, in their mid-sixties, stood looking at what was left of their two-bedroom, two-bath home in Homestead. It had no windows left, no roof. The furniture was scattered under an open sky. "I was in a daze," he said. "What are we going to do? Where do we go from here?"

The Morkills did better than Mary Cowan, who had telephoned her daughter the day before and told her not to worry. Mary had bought hurricane shutters and had them installed and in place before the storm hit. She had bought enough water and food to last until things returned to normal. But when Andrew hit, it shattered her house. The shutters held, but the roof blew off and then the walls collapsed on top of her. Mary Cowan's body was found two days later.

2

By Monday afternoon Andrew had left Miami behind, shattered and broken, dazed and bewildered, and was heading west across the Everglades. Its intensity died down as it passed over the state but renewed once it hit the Gulf of Mexico and a new fuel supply of warm water, and it curved north toward Louisiana. In Miami Kate Hale, the director of Dade County's Office of Emergency Management—whom no one in Miami could have named before Andrew hit; indeed, few people even knew there was such an office—stepped outside and took a quick survey. She didn't have to spend much time looking around; the disaster was evident, the need for emergency management of some kind obvious. She reached the state division of emergency management and told them to "send us everything you've got." Don't waste time asking what we need, she told them; we need everything you've got and more. Just get started.

Governor Lawton Chiles replied to reporters' questions about the extent of the damage: "I don't know how you describe the devastation. It looks like a bomb went off." He sent two thousand National Guardsmen to help.

President Bush was in the middle of a campaign trip in the northeastern United States. He immediately ordered Air Force One to turn around and head to Florida. Stepping off the plane later that day, he announced that he had already declared Florida a disaster area and that help from FEMA, the Federal Emergency Management Agency, was on the way. He toured the devastation and his heart went out to the people he saw. He grasped their hands and promised that the federal government would not desert them; it would bear the "full cost" of rebuilding public structures, he said. When he saw the private homes that had been totally destroyed, some without insurance because the people had put every cent they could scrape together into buying the homes, he promised that the federal government would reimburse the uninsured for their losses.

He visited Homestead Air Force Base and groaned aloud. The base had been not only the linchpin of our southeast defense posture, it had been the economic base of south Dade County. "The air base will be rebuilt," he promised, as people held out their hands to him and cried (as the *Herald* reported): "Help us, Mr. Bush. Help us, Mr. President."

During these stressful moments, he had forgotten that the federal government is not in the insurance business, and that if Washington started picking up the tab for the uninsured now, there would be little reason for anyone to bother paying for insurance in the future. He also seemed to forget that the decision of whether or not to rebuild Homestead Air Force Base—or any other military establishment—was Congress's, not his, to make; and that in fact Homestead was already on the list of military bases to be abandoned.

Well, it was an election year, and the people would be voting before these promises would have to be retracted, so perhaps he can be forgiven.

Three days later, Miami was still on its own. The *Miami Herald* on Friday morning blared a stark headline:

WE NEED HELP

As Dade County's hurricane relief effort neared collapse Thursday, President Bush ordered more than 2,000 airborne soldiers into the county to cope with what now is being called the most destructive natural disaster in U.S. history.

The military establishment had been waiting to help since before the storm hit. Major General John Heldstab is director of the army's military disaster-relief program. As the storm crossed the Atlantic the likelihood of it hitting somewhere in the United States had grown along with the storm's intensity and Heldstab's concern. He had begun planning how to help. Before the storm hit he had positioned millions of Meals Ready to Eat, the MREs made infamous during the Gulf War, and had transport planes ready to fly them wherever they were needed. He had tents, cots, blankets, water, electric generators, and bulldozers standing ready to be loaded and flown to the disaster area. A special task force was set up at the Army Operations Center, with banks of telephones open twenty-four hours a day to field requests for help and to coordinate relief efforts. Task-force members stood waiting for the phones to ring.

And they waited, and waited.

No one called.

Kate Hale started by calling the proper numbers, going through the communications route that had been established in advance. She called the state emergency management people, she called the governor: "We need food, we need water, we need people. If we don't get more food into the south end in a very short period of time, we are going to have many more casualties." She was told help was on the way. But it wasn't.

Instead, there was "bitter feuding" among the agencies that held shared responsibilities for relief. None of them seemed to know or care about the army's capabilities, no one thought to ask them for help. "Frustrated to the point of tears," the *Herald* reported, "Kate Hale said the relief project was on the brink of paralysis, a victim of incompetence and political games."

President Bush said it was a bad idea to play "the blame game." Help was on the way, he said. Governor Chiles, a political philosopher who felt strongly about federalism, thought the state could handle its own problems with a minimum of federal support. Certainly, he thought, the state should be in charge of the relief effort rather than just standing aside wringing its hands while the feds took over.

Meanwhile, people sat in muddy fields without a roof over their heads, without water to drink or food to eat, with nothing but—literally—a pot to piss in. Reporters drove through the ruined neighborhoods to take television pictures and were greeted with outstretched arms, people begging for help. The affluent moved out, to friends or hotels in the neighboring northern counties. The migrant workers, the elderly on small pensions, the support workers for the air force base—the normally invisible lower stratum of southern Dade society—became horribly visible because they had nowhere to go. They stayed and waited for someone to help them. And they waited.

"Where the hell is the cavalry on this one?" Kate Hale demanded, finally calling her own press conference in frustration and desperation. "We need food. We need water. We need people. For God's sake, where are they?"

3

Jimmy Carter, with the foresight he is not usually credited with, was the first president to decide that the country needed a federal agency designed to coordinate emergency relief if a disaster should strike anywhere in the country. In 1979 he created the Federal Emergency Management Agency, or FEMA, to plan for such emergencies, to organize a federal response, to ensure the availability of supplemental federal funds, and to train people in the types of work that would, without warning, become necessary and vital. President Ronald Reagan, with the blundering opacity he is now more and more frequently credited with, decided that the most likely disaster to hit us would be a nuclear attack by the Soviet Union. In light of the continual threat of hurricanes, earthquakes, floods, and tornadoes—disasters that have occurred from time to time and are guaranteed to do so again and again—along with the other possible natural disasters that might strike at any time, such as an outbreak of an epidemic disease, this was a remarkable decision to make. Apparently he thought so too, for he didn't seem to take FEMA seriously, and over the years it has received criticism from a wide variety of people. *Newsweek* reported that "it became a dumping ground for paying off low-level political debts; scandals demoralized its staff." When Hugo hit South Carolina in 1989, FEMA was nowhere to be seen, and Senator Hollings called it "the sorriest bunch of bureaucratic jackasses I've ever seen."

In October of that same year, when the World Series earthquake hit Oakland, FEMA did no better. Three years later the California state controller, still trying to pry federal funds loose, complained that "a natural disaster is being compounded by a man-made disaster in Washington." And San Francisco's Office of Emergency Services said much the same: "FEMA's response to everything is always frustrating. They have very cumbersome procedures."

In the rubble left behind by Andrew, FEMA seemed to have disappeared completely. "The question echoed through the debris Thursday," the *Herald* announced angrily. "If we can do it for Bangladesh, for the Philippines, for the Kurds of northern Iraq, why in God's name can't we deliver basic necessities of life to the ravaged population of our own Gold Coast?" The answer, the paper suggested, was simple: no one was in charge. Instead of delivering food, the helicopters that crisscrossed the skies of Homestead were delivering government officials, who came, looked, promised help, and then stepped back into their choppers and disappeared. Volunteers came by the dozens in their own cars and vans, bringing water and food and blankets, but were either turned away at roadblocks or directed to places their supplies were not needed. A typical story involves Sam Katz, of northern Dade County, who owns a chain of mattress stores. In the two days immediately following Andrew he set out on his own, loaded trucks with mattresses, spent his own money to buy food and baby supplies, but heard on the radio that individuals who did not live in the damaged areas were not being allowed access. So he called the Red Cross and asked how he could deliver his supplies. They didn't know. He called the United Way, and couldn't get through. "Three of us have been on the phone all day," he told the *Herald,* "and we've gotten nowhere. I've just been beating my head against the wall."

Within three days after last year's decision to help the Kurds, 8,300 troops were dispatched with 450 million dollars of relief aid. Three days after Andrew, a total of two thousand troops and five solitary airplanes had landed in Dade County, bringing in cots, roofing materials, and peanut butter. By day's end the supplies they had brought were still at the airport; no arrangements had been made to truck them into the besieged areas. That is not to say there were neither trucks nor drivers. There just weren't any directions for them; they had no clue as to where to go. Everyone stood around and waited.

Around the nation, people tried to help. In South Carolina, Georgetown County Administrator Gordon Hartwig, leader of a community that had been hit hard by Hugo and was grateful for the volunteers who poured in to help, announced that the county had five thousand dollars to send to Miami. "It's payback time," he said, and hordes of volunteers from South Carolina headed south to help rebuild. The Strohs Brewery Company shipped eighty thousand beer bottles filled with water; Kmart sent blankets and diapers; Amway sent cleaning products; Gerber shipped baby food. Police and firefighters around the nation began collection depots for food, clothing, and money. Cuban-American organizations jumped in to help their compatriots in their "capital," and citizens in Franklin, New Hampshire, organized a team of doctors and nurses.

Despite this goodwill, it was impossible to get the help to Miami. The first federal troops to arrive were stationed at the exits of the Florida Turnpike and given the job of stopping every car trying to get off. Motorists were asked for proof of residency, and if they couldn't supply this they were turned around and sent back. By trying to keep potential looters and rubbernecking ghouls out of the hard-hit neighborhoods, the federal troops, in their directives, failed to account for the possibility that they would also be turning away much-needed help. The Red Cross told doctors and nurses from New Hampshire that they weren't needed, that only people who had been through the Red Cross's own disaster-preparedness training would be allowed in. And where was FEMA?

Where, indeed? The only evidence I saw of their involvement was an announcement for an 800-number for "those who need help." That sounded good, I thought. We had no food left after a couple of days, and no stores nearby were open. We still had no safe drinking water and no way to boil it. We had some beer and Diet Coke left, and were not in any actual danger, and I knew that others were much worse off than we were, but still, I thought, what the hell? I picked up the phone to dial. The phone didn't work.

All through the day, phone service came and went sporadically. Whenever I could get a dial tone I called the number, but always got a busy signal. Eventually I gave up. Later I learned that my brother-in-law in Connecticut called the number, which had evidently been advertised up there as a way to locate missing victims. He got a recorded message telling him to call back in a few days; if it was an emergency he should wait for an operator. He waited, listened to recorded music for a few minutes, and then got a busy signal.

The head of the Red Cross, Elizabeth Dole, appeared on *Larry*

King Live! and told the world that the situation was serious but well in hand, that aid was pouring in and people were being taken care of. This was three days after the storm hit. It was not exactly true. Nearly a week later the Red Cross still didn't know what to tell people calling to donate food, clothing, and electric generators: "We get a call, we take a message, we give it to somebody who gives it to somebody else. We have no idea what happens to it. The whole place is being run by senior citizens and college kids."

"There is no point getting into blame," President Bush said. "I don't want one single federal official trying to be in the blame-assignment business."

Minutes later White House spokesman Marlin Fitzwater said it was all Governor Chiles's fault. And Chiles smiled bravely and said, "Every day things look considerably better. We know we've got enough food. We know we've got enough volunteers. . . ." And FEMA said nothing.

"Enough is enough," Kate Hale said. "Quit playing like a bunch of kids. We have people without water, people without food, babies without diapers and formula. . . . We're about ready to drop, and reinforcements are not coming in fast enough."

Replying for the president, U.S. Transportation Secretary Andrew Card told her that help was on its way although delayed by Governor Chiles's refusal to ask for it; he had finally asked two hours before and it was already on the move. Chiles replied that he had asked for help verbally; "We didn't have the time to write written requests and invitations," his chief of staff said. The lieutenant governor blamed "chains of command." Kate Hale said, "Apparently this whole thing is dependent on little pieces of paper."

Eventually the pieces of paper were written, signed, and delivered. Ten days after Andrew there were more than fifteen thousand soldiers and marines in Dade County, restoring order, directing supplies, putting up tent cities, helping people clear out their driveways, and pulling fallen trees off the streets. Another ten thousand were on the way, with MREs and medicine, food and water, baby formula and diapers.

Six months later Henry Cisneros, secretary of Housing and Urban Development, who had been put in charge of federal hurricane relief, issued a statement: Unless many additional hundreds of millions of dollars in federal aid was received, south Dade County would turn into

> a vast tangle of poor communities, uncoordinated and haphazardly constructed, home to the poorest, without economic function, and one of America's most troubled pieces of geography.... Neighborhood after neighborhood remains unrepaired. . . . There are endless piles of debris. . . . Social services are either closed or stretched to the limit. . . . Shopping malls and retail centers are boarded up and empty. . . . Homestead Air Force Base remains a tangled mess of twisted hangars, flattened buildings, and empty homes.

4

Richard A. Frank, administrator, National Oceanic and Atmospheric Administration:

> Since the Galveston disaster [in 1900] the United States has put men on the moon, orbited satellites to forecast the weather, and invented that miracle of modern civilization, the pop-top can. One might assume that our technological ingenuity has reduced or eliminated the risk of losing substantial numbers of lives in a hurricane. That is, however, not true. In fact, the hurricane peril has significantly increased.
> More Americans are at risk today from a major hurricane than at the turn of the century.

Is this true? Absolutely. We have to recognize that despite all our victories over nature—the gathering of the hidden energy of atomic nuclei into gigantic bombs, the abolition of smallpox, a successful vaccine against polio, airplanes that can fly faster than sound anywhere in the world—we are not yet masters of nature. There are things we simply cannot do, things we just do not know enough about to figure out.

In 1954 three successive hurricanes struck the eastern seaboard, raging from South Carolina through New York and up into Massachusetts and Maine; in 1955 two more hit the Mid-Atlantic states. More than four hundred people were killed, and Congress leapt into action by commissioning a five-year accelerated research program, a "Manhattan Project" to master hurricanes as we had so recently mastered the atomic nucleus. Aircraft reconnaissance was expanded, radar surveillance was added, and soon satellites were looking down on the earliest development of tropical storms.

What was the result? Although hurricane forecasting improved at first, results soon leveled off. And the hurricanes continued to come. Betsy hit New Orleans in 1965 and caused $1.4 billion damage, Camille hit the same area again in 1969 and killed 258 people, and several storms killed another 120 people in 1972. Congress's new hurricane project had begun in 1956; in the next twenty years there were an average of 50 percent more deaths per year attributed to hurricanes than in the previous twenty.

This occurred simply because more people had moved into the eastern coastal regions, and despite many scientific advances, no technological improvements have sharpened our ability to deflect or modify hurricanes. The most we have done is learn very little about how to predict when, where, and how forcefully they will hit. And the ability to predict may not always be helpful. As Richard Frank went on to say,

> The current state of the art in the evacuation field is not good. Most locales have no information at all about flooding under storm conditions. Most information about evacuation routes is out of date or otherwise insufficient. Old evacuation plans which do not reflect the boom in coastal populations may be more dangerous than no evacuation plan at all. If evacuation routes become overcrowded, we might lose more lives by exposing people to the risk of drowning in their automobiles than if they had stayed home.

Frank wrote that in 1980; by 1993 nothing had changed. On February 28 *The Miami Herald* reported the results of a local survey which indicated that two-thirds of the south Dade County population was planning to flee the next hurricane. The expected result was gloomy: The mass evacuation could be a disaster all by itself, experts warned.

"One of our major concerns is that too many people will try to leave the area," said Bob Sheets, director of the National Hurricane Center in Coral Gables. "We just don't have the road system to get all these people out. If that many people try to leave, there will be no way to evacuate the Keys, Key Biscayne, Miami Beach or Hallandale."

The evacuation would "create gridlock, stranding people on the road as a storm approaches. To properly evacuate that many people would require more than a three-day warning."

The sad fact is that with all our scientific advances, satellites and reconnaissance aircraft, we cannot give even a three-day warning of an approaching storm.

5

The world we inhabit is not a loving, beneficient place. Life in general, and our own lives in particular, consists of a continuing battle for survival against other forms of life that threaten to consume us —we no longer worry about wolves and tigers, but about the HIV virus, the malaria plasmodium, the syphilis spirochete—and against nonliving paroxysms of nature. Sixty-five million years ago an asteroid fell from the skies, and half the living species on earth were wiped out. Twenty thousand years ago the ice caps began to spread, and the northern hemisphere was crushed under a two-mile-thick mountain of ice stretching from the North Pole down to the latitude of New York-Chicago-Moscow. Three hundred and fifty years ago the Black Death came rolling in from the east, and two thirds of all people living in Europe died.

The world we live in is subject to these horrors, and to others that occur and reoccur several times within everyone's life span. Earthquakes and volcanoes occur every year somewhere on earth, once in a while causing great damage to humanity. Avalanches and mud slides, tornadoes and floods, bedevil us from time to time. But of all the destructive forces in nature's arsenal, the hurricane is the most destructive of all. Until recently there was nothing to do about it. We had no weapons with which to fight back and we could not even flee the field, for the fields on which the hurricanes attack are the coastal regions of the continents, and for a variety of reasons these regions are where our civilization has taken root and must remain. Thus we plowed our fields and built our cities, and when the storms came thrashing out of the sea we huddled down as best we could, prayed to whatever gods may be, and waited for the winds to pass.

The situation today is somewhat better. Not a whole lot better, but somewhat. Unable to control the monster, not yet wise enough to seed it to extinction or steer it along a path of our choosing, we can at least see it coming, if only from a short distance away.

Our knowledge began when Ben Franklin realized that storms travel from place to place, and when the invention of telegraphy and radio enabled us to track them as they came, from ships at sea and from town to town. With the advent of the airplane, radar, and satellites, we can spot developing storms and follow them across the oceans. With adequate warning we can avoid the loss of life, if not the loss of property. Still, even with all this technology, we can not yet give adequate warning, because a hurricane is not a well-modulated system with linear, calculable behavior. Rather it is a conflagration of wind and impulse that surges, swirls, and bounces about in a hodgepodge of vertical vortices and inscrutable sinks, that can dart right or left at the last moment. It can suck up a burst of energy just before landfall, so that it erupts like a bomb, or can whimper ashore, loll about for a while, and then fade away like a Cheshire cat, leaving behind nothing but a silly grin on the faces of those fearful people who bought out the supermarket and boarded up their houses for what turned out to be nothing, nothing at all.

If hurricanes moved in straight lines there would be no problem in predicting where each one would hit. You would need only two position fixes, and its future path would be certain. Or, if it moved along any well-behaved curve, its future path could be predicted if more position points were obtained. With satellite and airplane observations we can get as many fixes as we need; the data could be fit to a curve and the curve extrapolated to show the storm's future path. But it does not move in well-behaved curves.

If hurricanes followed the impulse of a well-defined steering current, we could predict their movements. But the steering currents are not well defined, are subject to fluctuations from other weather systems, and interact with the storm itself in manners that can be as little predicted as they are currently understood. In sum, while

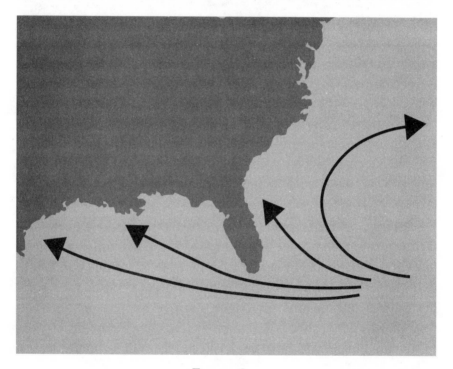

FIGURE 8-1

the average tracks of hurricanes can be easily represented (see Figure 8-1), the path of any actual hurricane is variable to an infinite degree. For example, Chloe in 1967 curved around and missed the Florida coast, and when it headed out to sea all the forecasts said the danger was over. But then it suddenly stopped and turned around and came back to clip the North Carolina coast before curving down toward Florida and around again out to sea, where it finally died (see Figure 8-2).

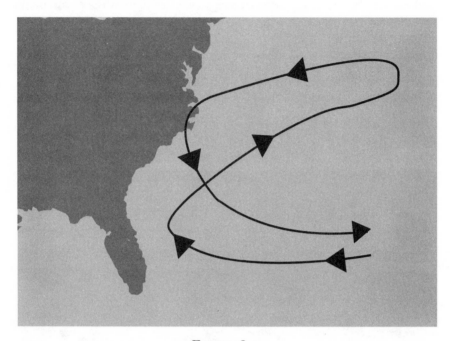

FIGURE 8-2

Presently, a forecast made twenty-four hours before a hurricane is expected to hit land can locate the position of landfall only within three hundred miles, at best. Since the area of serious damage is likely to encompass perhaps a fifty-mile diameter for the most serious storms, this means that more than 80 percent of the area warned will not be affected. If mass evacuations are made, 80 percent of the people will have been rousted from their homes for no purpose.

So what's to be done? Wait longer before giving the evacuation signal, so the position of impact can be better determined? But already the twelve-hour warning time is insufficient; our cities are not planned for mass evacuations. Highways are not built wide enough to handle even normal rush-hour traffic, let alone a wholesale exodus. In Miami, as Bob Sheets explained, we would need three days to move people out of harm's way. In the city of Miami Beach there are 100,000 people whose only access to the mainland is by four causeways, the largest having two lanes leading off the island and the smaller ones being more like neighborhood streets. All of them feed directly into I-95, the main artery leaving Miami proper—which can't even handle everyday traffic without clogging. This situation is typical for coastal communities. Can you imagine the horror of trying to evacuate Long Island or Cape Cod, and the political repercussions two days later when the hurricane swerved and missed land completely?

As Andrew approached, the best we could do was make educated guesses as to where it would go. The storm eventually hit Miami in the early hours of Monday, August 27. On the preceding Friday, the best estimate was that by Monday Andrew would still be east of the Bahamas. On Sunday morning, twelve hours before landfall, the odds were still only 40 percent that it would hit Miami, with a 20 percent chance of it hitting Key West instead (120 miles south), or a 6 percent chance that it would veer north and hit Vero Beach (120 miles north). It costs $1 million to evacuate each mile of coastline, and so the forecasters were desperately trying to narrow down the possible margin of error, but this was the best they could do. The resulting evacuation was the largest in United States history, even though only people living near the coast, in regions likely to be flooded, were told to get out. My father in particular was angry; he had to endure a lot of laughter and joshing when he returned to his condo and faced all the people who had stayed there and had watched on television as the storm missed them and instead hit exactly where he had fled to.

6

In the 1960s and 1970s significant progress was made in understanding the physical basis of hurricanes. With each new degree of understanding came a concomitant advance in the ability to predict where and how hard the storm would hit. It seemed then that soon we would be able at least to predict the path the storm would take, if not to regulate it. But as often happens in science, this surge in understanding reached a plateau. We seem now to be dangling along the edge of a new surge, but have not quite reached it. When Andrew came sailing along toward Miami there were computer models to analyze and predict its path; but what we had were *models*, rather than *a model*. Each model was an amalgam of different facts and assumptions, and each gave a different prediction. One of the models did in fact predict its path accurately, but there was no way to know in advance that this answer was the right one and the others were wrong. The model that gave the right answer for Andrew has given wrong answers for other storms, further delaying any sense of certainty. We don't even understand things after the fact; Hurricane Celia in 1970 dropped its central pressure from 988 millibars to 943 millibars in its last fourteen hours before hitting, leading to an explosive increase in wind speeds and subsequent devastation, but no one knows why. In 1981 Robert H. Simpson, co-creator of the Saffir-Simpson hurricane scale, wrote:

> The explosive development of disturbances and rapid growth of hurricanes into extreme events are currently beyond the reach of operational models and remain unresolved problems. These unresolved problems are the more important because significant changes in strength or size of the hurricane strongly influence the height of peak storm surges, the extent of coastal inundation, and the requirement for evacuation.

This statement is still true today.

7

The first serious attempts to predict the path of hurricanes began in the 1940s, as part of the war effort, with the realization that hurricanes are pushed along by external steering currents rather than wandering about of their own volition. It was natural at the time to try to correlate the storms' movements with the winds observed at the surface, but these attempts all failed. The next step came with our explosive exploration of the upper atmosphere, as high-altitude bombers and fighters began to move above ten thousand, fifteen thousand, and twenty thousand feet to fight the war. Scientists began to learn more about the winds at all levels, and about how distinct they were. Some workers suggested that the upper-level winds might be more important than those at ground level for hurricane development and steering, and the next round of movement predictions focused on these winds. And these predictions failed just as miserably as those which had come before.

Eventually it was realized that the hurricane is a three-dimensional creature and must be followed through all three dimensions; not only along the X-Y plane of the earth's surface or at high altitudes, but vertically, all the way from the bottom to the top. This seems obvious now, but it was hard to realize because it was impossible to actualize: three-dimensional calculations involving variable wind speeds and directions along the vertical axis were too complex to be carried out, even in approximations, and so if the workers at that time were to accept this conclusion it would have meant an end to all their attempts at prediction. Since meteorologists are human (although barely), they found it hard to accept this conclusion until there was a way out. Luckily, the way out came at right about this time, with the invention of the computer.

Using computers they were able to at least attempt to set up and solve equations relating the structure of the hurricane to the dynamics of the steering current at all altitudes. But unfortunately that was the most they could do: attempt it. The equations turned out to be too complex for direct solutions even with the most advanced computers, and that is still true today. In order to make any progress at all it became necessary to structure the equations so they bracketed parts of the problem and left other problems outside, to be approximated or estimated only roughly. The models fell into three classes: one dealt mainly with the structure of the hurricane and made only a rough approximation to the steering currents, another dealt more realistically with the steering currents but used overly simplistic hurricane structures, and a third class dealt with historical climatic relationships and their effect on hurricane movement.

The first meaningful model was known as NHC-64, since it was made at the National Hurricane Center in Miami in 1964. It analyzed atmospheric circulation at three different altitudes from measurements made on past hurricanes, and used these to compare with measurements of a current hurricane to predict its future motion. In the late 1960s the United States Navy developed a four-altitude model, and featured successive error corrections for movement in each twelve-hour period. About the same time NASA developed a model called HURRAN (Hurricane Analog), which was capable of quickly analyzing all hurricane tracks within the past hundred years and then calculating the most probable track for a current hurricane, given its position, direction, and speed, by comparing these parameters to those that occurred at the same place and time of year in the past.

In 1969 the first truly three-dimensional model failed because of a newly discovered complication: asymmetries in circulation, which seemed to be correlated with intensification. This discovery made the prediction of futures in any given hurricane immensely more problematic. Improvements in computers since then allow quicker calculations involving more and more parameters, but even so our technology is not yet up to the immensity of the problem. In the mid-1970s the National Weather Service began developing the first truly modern three-dimensional model, with grid resolutions of sixty kilometers. (The grid resolution is a measure of how tightly measurements and calculations are made. Ideally one would want a resolution smaller than the operational variations in a hurricane. If we could get down to one-meter resolution we would know everything that was going on inside the storm. Current technology is nowhere near this possibility.) By 1980 this model, NMC-Hovermale, was useful mainly for showing our limitations; it could not predict accurately the development or future direction of current hurricanes.

In observing a well-behaved hurricane—one that develops without sudden bursts of growth, and that moves along in a smooth parabolic curve—the models fit it and predict it quite well. But when the hurricane explodes within hours into a more destructive storm or swerves without warning to a new track, as so many of them do, the models simply fail: they cannot predict such behavior. Even with a well-behaved storm it is impossible to predict anything, since the storm might stop being well behaved without warning.

It has now been thirty years since scientists first started working on sophisticated computer models of hurricanes, and the error in predicting the path of any given storm has been reduced by just about 14 percent. This is not a happy result; it is so small as to be nearly meaningless. In 1990 a new model developed in Yugoslavia and being run at the National Meteorological Center in Maryland showed some promise. The model, called ETA—for no good reason that I can fathom—caused a happy stir when it was used to follow Hugo in September of that year. While all other models indicated that Hugo would dissipate after hitting the coast, thereafter stooging along the coastline, ETA called for it to head inland, toward West Virginia. When Hugo did that, people began to take notice.

ETA concentrates a lot of computing time on analyzing rainfall, more than other models do. And it is when the hurricane dumps its load of water that the latent heat is released, so the rainfall patterns are intimately related to the heat pump that drives the circulation. The hope was that ETA was able to simulate this process more realistically than other models, and therefore would be a more accurate predictor of hurricane patterns.

This hope has not been validated, at least in the sense that ETA has not run other models out of the game. As I write, in the spring of 1993, the gallimaufry of computer models seems to be expanding faster than the population of Asia. Just glancing through the recent scientific literature, I come across the Quasi-Lagrangian Model (QLM) of the National Meteorological Center (which is the model used by the National Hurricane Center for making operational forecasts); the National Hurricane Center's own NHC83 model; the moveable fine-mesh model (MFM); a model based on climate and persistence of motion (CLIPER); the Quasi-Lagrangian nested-grid model (QNGM); and a host of others, whose acronyms are ECMWF, GFDL, NMC, and VICBAR.

All of these use different assumptions about the parameters involved, concentrate on different aspects of hurricane structure and growth (such as evaporation or precipitation), and come out with different answers. That is why, as mentioned in Chapter 3, when the scientists gathered at the Rosenstiel School in Miami to discuss Andrew, we were told that a model had correctly predicted its path —but that all the other models had disagreed, and the one that worked for Andrew had failed for Hugo, and the one that worked for Hugo had failed for Andrew, et cetera, et cetera, and so forth . . . As Dr. Rainer Bleck, head of the school's Division of Meteorology and Physical Oceanography, said with a sad sigh, "It's very complicated."

The current state of the art was summed up exceptionally well by Dr. Mukut Mathur of the National Weather Service, in an article in 1991 introducing the QLM:

> The operational numerical prediction of a tropical cyclone's intensity and motion is very challenging. First, the initial structure of a storm cannot be resolved because of lack of observations near a storm's center. Furthermore, these storms often form in oceanic regions where data are so sparse that even the large scales of motion are not well analyzed. A model initial state based on such an analysis may have a storm structure and steering current that are quite different than those in the real atmosphere. Another difficulty arises due to the model configuration. Very small- (convective) to large- (planetary) scale motions determine the structure and track of a storm; a model whose domain extends over several thousand kilome-

ters and uses a fine horizontal grid spacing of a few kilometers is needed. Currently, it is feasible computationally only to use a small limited area with a grid spacing of tens of kilometers.

He is saying that there are problems both with the models and with the data. Kerry Emanuel, a meteorologist and hurricane specialist at MIT, thinks the problem lies more in the data than in the models. He suggests that the biggest uncertainty is in the measured conditions within the storm; the way the models work is that they take the initial measurements and then calculate what they might evolve into. But the initial measurements, like any experimental data, have errors associated with them; as the computational process proceeds, these errors multiply and magnify. The key to successful prediction, Emanuel thinks, is not so much with improving the models as with improving the precision of the data: the smaller the initial errors, the smaller they will grow during the modeling process. The major uncertainties, he says, reside in the sampling of each hurricane: "The conditions are undersampled, even with weather satellites."

8

The advances made in sampling storms since the congressional initiative of 1956 have been truly enormous, and it is a great disappointment that they have not resulted in a better ability to at least understand and predict the future behavior of hurricanes, let alone to modify them. Radar was the first of these, advancing to become a magical eye from an uncertain start as a weapon of war. In fact, the first time radar was used to combat the weather came about entirely by accident.

In 1937 a team of scientists in Great Britain was trying to make a radar set small enough to be carried in an airplane and sensitive enough to find other airplanes in the night skies. They had not yet succeeded, but on August 17, while conducting an unsuccessful test for air-to-air detection, they unexpectedly found that echoes were obtained from ships several miles away. This was a total surprise, because in previous tests the returns from the sea itself had swamped anything that might be seen from ships. But the new set they were testing used a shorter wavelength than the older ones, and this serendipitously avoided the sea-surface returns.

There was more serendipity to come. In that same year the Royal Navy had prepared a fleet exercise designed to test the efficiency with which aircraft could detect ships at sea. The exercise had been set up before the discovery that radar might be useful; at that time it was not yet clear that airplanes would be of any use at all.

Units of the British fleet were sent into a wide area stretching from the Straits of Dover to Cromarty Firth on the northeast coast of Scotland. Forty-eight aircraft of Coastal Command commenced the search at dawn on September 4. In the air also was one old clunker of an outmoded type, an Avro Anson, not officially scheduled to take part, carrying the first airborne radar set designed by Ted "Taffy" Bowen. As he later said, the opportunity—coming just a couple of weeks after he had accidentally discovered that he could see ships by radar—was just too good to miss.

Within a few hours the weather began to dirty up. It quickly deteriorated below the level at which safe flying could be performed, and the exercise was cancelled. But since Bowen's old clunker was not an official part of the RAF and he had not bothered to tell anybody what he was doing, Bowen flew on, slipping between the clouds and, one by one, the ships of the Royal Navy began to show up on his primitive radarscope. They steamed on serenely, undiscovered, they thought, beneath the gray haze and towering thunderclouds, when without warning the Anson whirled down out of those clouds and sped across them. Happy and satisfied, Bowen and his two pilots turned and headed for home.

But the weather had become really nasty. They climbed slowly into a solid wall of gray haze intermittently darkened with black clouds of turbulent water vapor, somewhere behind which was the craggy shoreline of England. The pilots flew on instruments while Bowen crouched over the cathode screen. Ahead of them the radar beam poked through the opaque clouds and returned echoes of nothing but choppy water—and then suddenly he saw something different on the screen. A distinct difference began to show up, and with joy Bowen recognized the coastline clearly enough to give directions home.

Meantime the fleet at sea had radioed that they had been discovered by a lone aircraft, and whose the hell was it? Coastal Command checked with all its squadrons and found that all had returned upon the cancellation order, and that none had come anywhere near the fleet. Finally the old Avro Anson landed safely, and Bowen notified the RAF that a civilian aircraft equipped with radar had been able to fly out to sea when Coastal Command was grounded by the weather. The craft found the Navy by itself, he reported, and returned to base guided by radar navigation through weather that birds were watching from the ground. They all realized a new phase in naval warfare had been suddenly introduced. So too had a new phase in the scientific investigation of hurricanes.

In this first use radar showed a behavior exactly opposite to what would be needed to study hurricanes: Its beams penetrated the clouds and showed the land beneath it, which is fine for navigation but useless for weather reconnaissance. To study how a hurricane is developing, it is not helpful to have something that will penetrate the storm and see what is on the other side. What is required is something that will see the storm itself. That means an electromagnetic beam of some kind that will interact with the storm system rather than pass through it. At first glance radar does not seem to be the desired weapon.

But remembering the greenhouse effect, we know that radiation of one wavelength—say, visible light—will pass through carbon dioxide, while radiation of a different wavelength—such as infrared—will not. Radar consists of an electromagnetic beam that is absorbed and reemitted by an object, with the reemitted beam returning to a detector, thus revealing the presence of the object. The electromagnetic beam used in a radar system can have any wavelength the system is capable of producing. That first airborne set used by Taffy Bowen to find his way home by peering through the stormy weather had a wavelength of 1.2 meters; other wavelengths were soon in use by the military for various purposes, and eventually it was found that some wavelengths would interact with the water droplets in clouds rather than penetrate them. And thus clouds, and storms, became visible to the radar eye.

This is seen on television nightly when the weatherman shows the radar map, which illustrates the cloud distribution around the region. By the 1960s we were constructing a national network of weather radars along the eastern and Gulf coasts; by the end of the 1970s the network was complete and in operation, providing a constant view of any cloud buildup within 240 kilometers of the coasts (the range limit being imposed by the curvature of the earth). These radars were sensitive enough to show not only that there was a cloud system out there, but how much water was precipitating from it.

Nevertheless, there were still two problems remaining. First was the limited range of the coastal radar system, which couldn't see much more than a couple of hundred miles out to sea. This meant that the sites of origin of each year's hurricanes were out of range: the system could see the storms as they approached land, but could not measure their beginning and development from off the coast of Africa through the Atlantic Ocean. Second was radar's inability to measure wind speed, which, as we have already seen, is what defines the strength of a hurricane. So even when the system saw a storm approaching, it was unable to give a reasonable estimate of how bad it was going to be.

These deficiencies have now been remedied. Wind speed can be measured by radar systems incorporating the Doppler effect, which is easiest to understand in the case of sound waves.

Sound is generated by something vibrating within a medium, the usual medium (for us) being air. When you pick on a guitar string it vibrates, and this vibrating motion is imparted to the surrounding air molecules and passes through them to our ears and is sensed as *sound*. The pitch of the note produced is dependent on the length of the guitar string, which vibrates in a wave motion; the length of the wave decides the pitch of the sound, a shorter wavelength producing a higher pitch, a longer wavelength producing a lower pitch.

Now imagine a guitar-picker who is moving away from you (a circumstance devoutly to be wished). Since the vibrating string is moving away from you during the time that it is producing a wave, the wave is stretched out, and this produces a longer wavelength, which produces a lower pitch. (This is relative to you, of course; a listener moving with the guitar player would sense no change in wavelength and so no change in pitch.) The result is that as the sound source moves away from you, a constant pitch (as plucked) results in a continually lowering pitch (as heard). And the opposite is also true: a sound source moving toward you produces a continually rising pitch. This is the reason for the change in the wail of an ambulance siren, rising as it comes up behind you and falling as it passes by and roars away.

This is the Doppler effect, and holds true for all wave motion, including that involved in the generation of electromagnetic waves. It is what causes the Cosmological Red Shift we see in stars from distant galaxies. The wavelengths of the light they emit is longer than the same light from nearby stars, which tells us that these distant stars are all moving away from us. (In visible light the wavelength controls the color just as in sound it controls the pitch; a longer wavelength is a shift toward red, a shorter is a shift toward blue.) This is how we know the universe is expanding.

And this is how radar can tell us how strong a hurricane is, as in Figure 8-3.

The radar beam along *a* hits the water droplets within the whirling clouds that are spinning toward the radar set, while the beam along *b* hits clouds that are spinning away from the set. Therefore at *a* the reemitted radar beam has a shorter wavelength (as seen by the observing radar set) than does the beam coming back from the clouds along *b*. The difference in these wavelengths is a function of the speed at which the clouds are spinning, i.e., the wind speed of the hurricane.

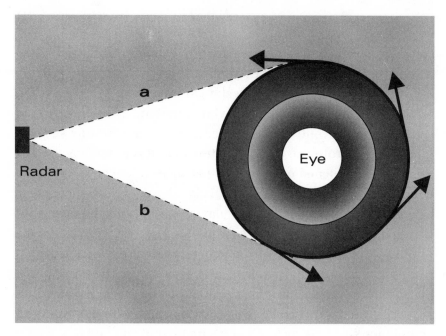

FIGURE 8-3

The other major advance in technology during these years was the geosynchronous satellite, which enters an orbit with a speed that matches the rotation of the earth. As it circles the earth it appears to an observer on the ground to be perfectly still in the sky. It sits over a chosen point and looks down on it, so that it can be positioned over the hurricane pathway in the Atlantic and watch continuously as storms develop. With a fleet of these satellites providing ongoing surveillance over the Atlantic and the Gulf of Mexico, we can see the birth and growth of hurricanes from beginning to end.

Unfortunately our vision is limited to only the outside of hurricanes, as satellites monitor with television. Infrared sensors can help measure the temperature, from which scientists can deduce something about the intensity of the internal processes, but the vast intricacies of the storm's innards remain invisible. It is as if, in an attempt to understand human growth, we had a videotape of a person from birth to death. It would be valuable, but what we really would want to know is what is going on inside the skin, even at the molecular level. In a hurricane the very thickness and opaqueness of the clouds generated serve to hide what is going on inside from the geosynchronous satellites and the coastal stations. What is desired in hurricane surveillance is something analogous to X rays, which can see inside a patient's body.

Although easy enough to discuss, this is quite difficult to accomplish. X rays work because the living body is composed of tissues of varying density and atomic structure. So it was possible to find a beam of radiation with a wavelength that would penetrate skin and soft tissue, for example, but would not penetrate bone. Skin and soft tissue are made primarily of carbon and hydrogen, while bone is made largely of calcium, a much heavier atom with a more densely packed electronic structure.

But hurricanes do not have such a heterogeneous structure. To have a better understanding of the storms' layers scientists should first measure the wind speed of the outer ring of clouds, which can be done with Doppler radar; and second, peer through that ring and measure, step by step, the wind speeds and pressure gradient along successively deeper layers inside the storm. But the inner layers are made of exactly the same stuff as the outer layer, and so a beam that penetrates the outer layer will continue to penetrate the inner layers and will simply come out the other side as if nothing had been in its way. It will tell us nothing about the structure of the storm since it did not interact with it in any way.

The situation is not hopeless, because it is possible to construct an electromagnetic beam that partially interacts and partially penetrates, but so far such an apparatus has been only partially successful. By using microwaves we can see through the cloud cover to some extent, and just recently a new system of Doppler radars, known as NEXRAD (NEXt generation of RADar), has been installed around the Atlantic and Gulf states, providing a new observational capability. The system, costing $850 million, overran its cost estimates in 1992, but this seems to have been worked out now, and all 116 radars are expected to be running by early next year.

And NEXRAD is only part of the story. In the coming few years the National Weather Service will be undergoing a $4.4 billion upgrade of its forecasting capabilities; it is literally exploding with new ground- and space-based technology. Another $118 million is being spent on Automated Surface Observing Systems, which will continuously monitor atmospheric pressure, temperature, winds, visibility, and precipitation—but only over the continental United States and out to sea for a distance of 240 kilometers. This will be extremely useful for observing the development of continental storms such as tornadoes and thunderstorms, but won't be of much use for hurricane warnings.

In addition, a new generation of computers will arrive at the National Meteorological Center in Camp Springs, Md, in 1995, with the installation of a Cray 90 supercomputer. Associated with it will be an "improved hurricane forecast model," as *Science* puts it. But the test of the pudding is in the eating, and judgment of just how "improved" it turns out to be will have to await an evaluation of its predictions when the next wave of hurricanes hits.

Those parts of the new system installed and operating have already proved their worth. Tornado and flash flood forecasting in the Midwest has clearly improved, with longer advance warnings and fewer false alarms but, as an old song lyric puts it, "Baby, we've still got a long way to go." The technologies listed above are ground-based or space-based, and between these two extremes is a massive hole. The space satellites can look down on the whole world, but can't see precisely enough into the interiors of hurricanes, while the NEXRAD Doppler radars can't reach far enough out from shore to watch the hurricanes as they develop. For the most precise information—which is what we need—we can only go to the same method the physician uses with the human body: he sticks a thermometer inside to find out what the temperature fluctuations are, since they are diagnostic of what takes place there. It is not as good as a CAT scan, it does not provide as much information, but often it provides enough for a meaningful diagnosis. Analogously, we send airplanes into the storms to measure the distribution of temperature, wind profiles, and velocities horizontally throughout the bands and vertically from top to bottom, and a host of other parameters. Unfortunately, although this information is tremendously valuable, it is not enough to provide a rigorous analysis. An obvious limitation is the inability to monitor a hurricane's development continuously with airplanes. In addition, the airplanes available can't do a complete job: the air force's C-130s and NOAAs PR-3Ds, which comprise our current staff of hurricane hunters, are prop-driven planes with operational ceilings of about ten kilometers. Our most recent models indicate that the topmost part of the hurricane's circulation—reaching up beyond eighteen kilometers—can have substantial effects on the storm's future motion. If it is not possible to get high enough to study those winds, it is going to be impossible to predict which way the storm is heading. Jet planes could do the job, but jets are expensive. And instead of plowing more money into our operational capabilities, we have been doing the opposite: in the past few years the reconnaissance budget has been drastically cut.

This has been done despite the fact that reconnaissance flights provide the most direct information needed for evacuation decisions. The hurricane's two most destructive aspects are, to repeat, storm surge and wind. The storm surge can be predicted for an oncoming storm with a computer model called (for obvious reasons) SLOSH. But to run this model, we need to know the central pressure of the storm, the radius of maximum wind, and the general wind-field characteristics, because the storm surge will be greatest for storms of lowest central pressure, but will vary if, for example, the wind speed drops off rapidly or gradually with distance from the eye.

A rather small variation in maximum wind speed, say about twenty miles per hour, could result in a difference of up to 200,000 people being told to evacuate. Satellite measurements can give information on wind speed and on the radius of maximum winds, but not as accurately as can an aircraft penetrating the storm; nor can the satellite eye penetrate and give precise information on the internal heterogeneities. Still, it seems that we simply can't afford to put another jet plane or two on the job. It may be difficult to understand why a nation with the world's largest jet-equipped air force, with more planes capable of flying anywhere under any weather conditions for the purpose of dropping bombs, a nation that is not at war and yet must keep this air force trained and operational for future emergencies — has any shortage of airplanes ready and able to search out these storms. Bending over backward as far as I can, I still find it hard to understand why these planes are kept on the ground while hurricanes cost us hundreds of lives and billions of dollars. I guess it's just part of the joy of having a bureaucratic government.

But as Kurt Vonnegut likes to say, so it goes. And so when Dr. Emanuel rightly points out that the models are not going to improve until there are more and better data to test and prime them with, and at the same time we're cutting back on aircraft reconnaissance, we have to acknowledge that our ability to predict just when and where and how hard the next hurricane is going to hit is not going to improve in the near future.

9

Lacking the capability to give accurate warning more than twelve hours in advance of just where a hurricane will hit and how bad the damage will be, and acknowledging the implausibility of evacuating most major population centers within that period of time, what is to be done?

In the early 1980s a lot of thought was given to the concept of vertical evacuation. The idea is to store people high above the storm surge, rather than to take them horizontally out of the area. It rests on the supposition that it is possible to build high-rise buildings strong enough to withstand both the surge and the winds of the hurricane, so that people can ride out the storm in the upper stories and then return to ground level when the storm is over.

When I was a child there were no high-rises in Miami; expert opinion said that you couldn't build a skyscraper there because—

well, no one knew why, really. Something to do with the ground, never precisely articulated. For years, decades, no one questioned that opinion. There was a similar situation at the University of Miami's Marine Laboratory when I joined it in 1966: no females were accepted as graduate students because all grad students had to do research at sea, and you could not allow women at sea because—well, no one knew why, and no one asked why not. And then a decade later someone finally did ask that question and found there was no sensible answer; today half the grad students are women. And when someone finally asked why they couldn't build a skyscraper in Miami, no one had an answer. So they built one, and it didn't fall down, so they started building a lot more. By the 1980s the authorities had identified more than forty buildings as suitable for vertical evacuation in case of a hurricane.

But in 1993 when Andrew struck there was no talk of using this idea; no one even mentioned it. Instead, people living in all the high-rises threatened by the hurricane—people like my father— were warned to get the hell out of their buildings.

This change of heart and loss of faith in vertical evacuation is curious. No one wants to talk on the record, but it's a victim partly of our greedy, crooked natures. It is certainly possible to build structures that will withstand hurricanes, but it is also possible to skimp on building materials and cheat on the building code, and simply to do poor work with cheap, unskilled labor. So what if this happens, and the years pass by and then a hurricane comes and you tell thousands of people to take refuge in these buildings, and they fail, they fall apart under the battering of the hurricane? What if these high-rises collapse and spill all these people into the turbulent hurricane waves? No one wants to take that responsibility, if such a scenario is at all possible. And it is possible.

You can design a building as perfectly as possible, with legislation that would enforce perfectly safe structures, and with building inspectors to make sure the buildings are built in compliance with those laws and in accord with those blueprints, but as Dorothy Parker once said, you can lead a horticulture but you can't make her think. And you can't make greedy, crooked constructors obey the law, and you can't make greedy, crooked building inspectors enforce it.

People began to realize this when Eloise hit Panama City Beach in 1975. After the storm, city officials sent structural engineers in to assess the damage. What they found was evidence not of how strong the hurricane was but of how weak human consciences are. Over and over again they found that buildings had not been built according to the blueprints, and somehow these faults had never been found by the building inspectors. One fourteen-story high-rise was built with only two thirds of the structural supports called for in the plans, and nearly half the concrete pilings had no concrete in them.

After Andrew we found similar crimes in Miami. Buildings constructed in the 1940s, 1950s, and 1960s were built well, sufficiently reinforced to withstand hurricanes. But in that time period only two big storms hit Florida, and as the sixties waned and the seventies and eighties rolled in, no storms came with them. Little by little the building standards were ignored; little by little, more and more, until the final wave of construction was producing houses and condos that simply could not stand up to the fury of the storm. Windows blew out and walls collapsed and roofs came off, and post-hurricane inspection showed that in many of the new buildings, the old regulations had simply been ignored. A University of Miami study claimed that gable-end trusses were not braced properly in many homes whose roofs had blown off, shingles had not been stapled properly, and ceilings had not been built to withstand a hurricane's heavy rains. Some buildings had less than the required number of anchoring tie beams, which led to the total collapse of the walls. The wooden frame houses had not been sealed properly, and moisture that had gotten in over the years had weakened them.

Immediately after Andrew several groups of Miami homeowners filed suit, accusing the Lennar Homes (Miami's largest home-construction firm) of failing to build their homes in compliance with the south Florida building code. Lennar denied the allegations and vowed to defend the lawsuits aggressively, but five months later the corporation settled out of court, agreeing to pay $2.4 million to the homeowners. In a statement, the company said it "remains firm in its belief that it was not responsible for damage caused by the extraordinary storm," but was settling for the usual reasons.

A second group is suing Walt Disney, which owned another major construction firm, for what they call "Mickey Mouse construction: walls that lacked steel and concrete reinforcement, roof trusses that were too weak for the job, improperly anchored or poorly installed, missing or undersized hurricane straps, and support posts that weren't anchored." Attorneys for the construction firm reply that the destruction of the buildings was not their fault, but was "directly attributable to the ferocity of the storm."

And perhaps they're right. For while buildings can be built to withstand a normal hurricane, what happens if at the last moment the storm builds up explosively to a category 5? Something of the sort happened with Andrew: no one is sure exactly how strong the winds actually were. In April of 1993 the argument was still raging, with meteorologists and wind engineers at odds. The meteorologists were saying that gusts hit 175 or even 200 miles per hour; the engineers claimed 140, max.

If the winds hit 175 to 200 miles per hour, the construction really cannot be blamed: no building can be guaranteed to withstand such a force. If the maximum winds were closer to 140 miles per hour, then construction could be faulted. But as far as vertical evacuation is concerned, it really doesn't matter. As long as there is the possibility of shoddy construction *or* of the hurricane escalating in force within the last few hours before landfall, you have to give up. For how can you tell people twenty-four hours in advance to take refuge in the upper stories of their condo, and then just a few hours in advance tell them that the storm has become a monster and they're not safe there and that they have to evacuate, get out of town with not enough time to make preparations? Especially since there is the very real possibility that such a flood of last-minute refugees would clog the arterial roads, and the people would be caught in their stalled cars by the racing floodwaters.

So, little by little, people have just stopped talking about vertical evacuation. It's an idea whose time has went.

10

Well, then, what is the answer?

(a) Modify and weaken hurricanes.

(b) Steer them away from vulnerable areas.

(c) Preparation by homeowners: buy solid homes, prepare with storm shutters, stock up on food and medicine, and ride it out.

(d) Preparation by local government: prepare well-built shelters and sufficient access roads for evacuation.

(e) All of the above.

(f) None of the above.

It's a trick question. The answer is (e) *and* (f). We've got to learn how to control the hurricanes, but we don't have this knowledge now and won't get it overnight, and even when we do, we'll still face the political and practical problems involved with any modification of nature on such a scale. As I've said, hurricanes are a necessary part of natural irrigation, bringing freshwater drinking supplies to much of the American and Asian continents, and we fool with them at our own risk.

Preparation by homeowners is necessary, as is preparation by the local government. People in coastal areas should be prepared to evacuate, and people farther inland should be ready to ride it out. Buildings should be built in accordance with proper construction codes, and residents should understand what survival through a hurricane involves.

But there is no one answer. There is no answer at all that does not involve knowledge and restraint beyond anything we've demonstrated so far. And, to be honest with ourselves, Andrew was the worst natural disaster ever to hit the United States, but it is not going to hold that record forever.

It hit south Dade County head-on. Dade County is a major population center, and that is why Andrew caused so much damage. But south Dade County is the least populated part of the county, with most of the buildings being single-family residences instead of large multidweller structures. If the storm had hit just a few miles farther north, the damage could have been infinitely greater. Instead of finding that the most recently built homes were constructed of faulty materials and with sloppy worksmanship, we may well have found that the most recently built condominiums and office high-rises had been similarly constructed. Instead of a single family being battered when their roof blew off, thousands may have been doomed when their giant apartment building fell apart.

Most of the people in my father's condo remained behind, ignoring pleas to evacuate. Afterward they congratulated themselves and joked about the people who left. But the storm ended up hitting forty miles south of them. Had it hit farther north, they would not be chuckling so heartily.

But it didn't. And since no storm like it has hit Florida for more than fifty years, it may seem safe to expect another fifty years before another monster comes along. When I moved to Miami in 1966 I was warned that hurricanes struck every seven years or so, but more than two decades then went by without one. Perhaps the climate has changed so that fewer hurricanes are generated nowadays, with Andrew just a rogue storm, the exception that proves the rule. Or have we merely gone through a transient phase, and are now returning to more normal frequencies of hurricane impact? The situation might become even worse because of the greenhouse effect and global warming, which could make hurricanes more frequent and vicious, allowing them to range farther afield, striking increasingly along the northern coastline as temperatures increase year by year.

Perhaps, and perhaps not. Certainly more hurricanes will be coming but, as Rodgers and Hart sang a few years ago, "Who knows where or when . . . ?"

9

FUTURE ENCOUNTERS

*In my head are many facts of which I wish
I was more certain I was sure.
Is a puzzlement!*

—from "A Puzzlement,"
by Richard Rodgers and Oscar Hammerstein II

1

The biggest danger threatening the future of our civilization is greenhouse warming. If this statement is true—and in my opinion it most certainly is—and if hurricanes are the greatest natural disaster ever to hit our country—which they certainly are—then one would expect a host of informed articles, books, and television programs about how hurricanes will develop in a greenhouse-warmed world: will there be more of them, will they be more destructive, will they affect more parts of the world?

But there has been no such host of articles, books, etc. Why not? The answer is easy: because at the present stage of our knowledge, any answer to that question would be premature. To illustrate why, we have to look at the subordinate questions:

Is the greenhouse effect for real?
Is it operating today?
Will it make the world warmer?
How much warmer, and how fast will temperatures rise?
Will it affect rainfall as well?
Will the effects be global or local?

Will a warmer world mean more hurricanes, or more ferocious hurricanes?

What else is likely to happen over the next few decades?

Beginning at the top, we know the answers rather well; as we progress toward the bottom they get vaguer and vaguer.

2

Is the greenhouse effect for real, and is it operating today? The answer is very definitely yes, but not necessarily in the way many people think.

The Second Law of Thermodynamics says that when a body gets warmer than its surroundings, energy will flow out of it. This can happen in various ways. The heat can be conducted out, as when ice cubes are put into a glass of warm water. In this case heat is conducted from the warm water to the cold ice cubes, the ice cubes get warmer and melt, and the water, as a result of its heat loss, gets colder. Another method of heat loss is by the emission of electromagnetic radiation, such as is the case with the sun; the hot sun radiates the energy, which makes life possible on earth, 100 million miles away.

It is this radiation of electromagnetic energy that concerns us now. The mechanism is simple. Atoms consist of nuclei surrounded by clouds of electrons. Although a detailed explanation lies solely within the mathematics of quantum theory, a pictorial representation explains the principles perfectly well. In this picture we have a central nucleus with electrons whirling around it in well-specified orbits. When the atom is heated, the electrons jump to higher orbits, in which they are unstable. As soon as they can, they drop down again to their ground-level orbits, and in doing so they emit a quantum burst of electromagnetic energy corresponding to the energy difference between the excited orbit and the ground-level. If the atom is heated only a small amount, the electrons will jump only to the nearest (lowest) orbit, so that when they drop down again the energy emitted will be small; if the atom is heated much hotter, the electrons will jump to farther orbits, and the resulting electromagnetic radiation will be more energetic.

The energy of the radiation is related to its wavelength: the greater the energy, the shorter the wavelength. Therefore when an object is heated very hot, the resulting emitted radiation is of short wavelength; when it is heated mildly, the radiation it emits is of longer wavelength. Since this general scheme is true for all atoms, the temperature-wavelength correspondence holds true no matter what the object itself is composed of.

Outside of a rigidly controlled laboratory experiment, any heated object will have electrons bouncing up to a variety of orbits; the emitted light will then cover a wide range of wavelengths. But it will still be true that the hotter the object, the shorter the wavelengths. Since it is "intuitively obvious" that the hotter the object the more intense its radiation (the brighter it will burn), two objects heated to different temperatures will emit radiation according to this scheme (see Figure 9-1).

Curve A is a hot object, curve B a much cooler one. For example, A might be the sun (at a surface temperature of about 15,000 degrees) and B might be the earth (at a temperature of about 20 degrees). In this case the sun's emitted radiation peaks at wavelengths of about 10^{-8} (0.00000001) cm, which just happens to be in the region of the electromagnetic spectrum that we call "visible light." (It didn't "just happen," of course; our eyes evolved an ability to sense that radiation because it's where most of the sun's energy comes off. Creatures evolving on a planet around a hotter star would probably evolve eyes that could "see" shorter wavelengths, if they evolved eyes at all. But this cannot be known for sure until another planet around another star is found.)

The earth—or any body at a temperature of about 20 degrees—emits much longer wavelengths, about 20 microns; we call this region of the spectrum "infrared." Incidentally, this is why infrared is often called "heat radiation"—because warm, living bodies are at about this temperature, and so that is what they emit. So in Vietnam

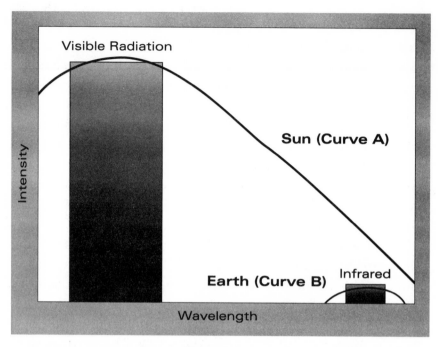

FIGURE 9-1

we devised infrared night scopes, which could "see" infrared, and so we could look in the dark and see any living creatures, like Vietcong. But the term *heat radiation* is a misnomer: *all* radiation is emitted by hot bodies and all hot bodies emit radiation. In fact the infrared is emitted by only moderately warm bodies; hotter things emit radiation at shorter wavelengths. (The night scopes, to be effective, had to be sensitive to wavelengths emitted by living bodies, but not to those emitted by plants or the earth itself; since living people are at a slightly different temperature than plants and dirt, this was possible.)

As explained in Chapter 2, the greenhouse effect arises because certain molecules are transparent to some wavelengths but opaque to others. Carbon dioxide and water vapor, among others, are greenhouse gases. They allow the sun's energy to come in through the atmosphere and warm the earth, and they trap the infrared radiation, which is then emitted from the warm earth. The effect is as if a blanket had been thrown around the earth: blankets do not provide any heat themselves, they just keep the heat emitted from the wrapped body from escaping. And just as you are warmer with a blanket wrapped around you than without one, so the earth is warmer with the greenhouse effect than it would be without it.

Drastically so, which we can see by looking at our nearest neighbors in space, whose temperatures look like this:

PLANET	DISTANCE FROM SUN (RELATIVE TO EARTH)	AVERAGE TEMPERATURE (°C)
Mercury	.3	~150°
Venus	.7	~450°
Earth	1.0	~20°
Mars	1.5	~−50°

The temperature in general is decreasing as distance from the sun increases, which seems reasonable, but it does not decrease in quite the right manner. First of all, Venus is hotter than Mercury's average temperature, although it's nearly twice as far away from the sun. Earth, just 30 percent farther away, is hundreds of degrees colder, yet precise calculations indicate that Venus should be just a bit warmer than the earth, and earth itself should be below freezing; that is, if the greenhouse effect is ignored.

It is the greenhouse effect that brings Venus's average temperature up beyond Mercury's, for the Venusian atmosphere contains about 95 percent carbon dioxide, while Mercury has no atmosphere at all and thus no greenhouse effect. Mars has such a weak atmosphere that its greenhouse effect is negligible; if earth was the same our temperature would be about 20 or 30 degrees colder; that is, below freezing. It is the 0.03 percent of carbon dioxide in our atmosphere that keeps our seas unfrozen and that makes the earth habitable; without it we would be a frozen, dead planet.

So the greenhouse effect is certainly for real, and is of inestimable value to us. The question that needs addressing, though, is whether we are increasing the carbon dioxide and thereby making the planet warmer.

3

Again the answer is clear: yes. The atmosphere has no horizontal borders; that is, it circulates rather freely completely around the globe. There is a bit of a holdup crossing the equator, but nothing of much significance for the issues we're discussing here. This simply means that the measure of carbon dioxide anywhere in the atmosphere will be typical for the entire world. (This is not absolutely true; for example, CO_2 diminishes in the spring, when plants are actively sucking it up and growing, and it increases in the winter, when their metabolism slows down. Since the seasons are not global [winter in one hemisphere corresponds to summer in the other], these variations are local, at least hemispheric, but they are also quite small compared with the global temporal variations we're talking about.)

Scientists have made a few measurements of carbon dioxide in tiny bubbles within ice cores drilled deep beneath the snowy surface of Greenland. These record the ambient air and its carbon dioxide content thousands of years ago, before there were enough people or any industrial activity to affect anything. Since the atmosphere circulates globally, we can reasonably take these data as indicative of the earth's atmosphere as a whole at that time. Beginning with the late eighteenth and early nineteenth centuries, carbon dioxide measurements became increasingly available, and in the twentieth century there is no lack of them. The picture they present is startlingly clear: for thousands of years the carbon dioxide content was rather constant at about 275 ppm (parts per million: 1 million liters of air would have 275 liters of CO_2), or 0.0275 percent. This geologic average stayed the same throughout our early history, but beginning roughly in the middle of the nineteenth century it began to climb. By 1900 it had topped 310 ppm, by 1960 it had reached 320, and it is now over 340 and climbing more steeply.

The curve of human population growth follows the same trajectory. Well, that's no great surprise and no coincidence: carbon dioxide levels mirror population growth since the beginning of the industrial revolution, when people began learning to use machines to take some of the drudgery out of life, and began moving around with the aid of the steam engine and later the internal-combustion engine, fabricating new materials and constructing new devices. The energy these advances required were derived from fossil fuels —coal, oil, gasoline—or are of their precursors, wood. No matter which is consumed, the process that gives energy can be represented by the simple equation $C + O_2 \rightarrow CO_2 + energy$, since the basic constituent of all these fuels is carbon, and the energy production process is its oxidation. So along with each unit of energy produced, a molecule of carbon dioxide is also produced.

The data are irrefutable; no matter the point of view of anyone who has studied the situation, all agree that the carbon dioxide content of the atmosphere is increasing. And all agree that carbon dioxide is an effective greenhouse gas. Moreover, another carbon-containing compound, methane (CH_4), is a more effective greenhouse gas and is increasing in the same manner as carbon dioxide, but even faster.

This was at first confusing, since methane is not associated with any of our human activities, at least not in such a way as to increase so rapidly. We do use it as a fuel—it is what is called "natural gas" —but most of us feel that leakage from mining sites could not explain its drastic increase, and when we burn it the reaction is $CH_4 + O_2 \rightarrow CO_2 + H_2O$, so that it changes to carbon dioxide, helping to increase the latter's concentration in the atmosphere rather than to increase its own.

The answer became clear eventually: methane is produced by some very ancient organisms on earth. Early in earth's history, before there were either plants or animals, the atmosphere had no free oxygen. A class of anaerobic bacteria evolved, bacteria that can exist without oxygen and that get their energy by reducing carbon instead of oxidizing it. The pertinent reaction is $C + 2H_2 \rightarrow CH_4 + energy$. With the advent of an oxidizing atmosphere these creatures had to hide, for oxygen is poison to them. They found an ecological niche in a few places on this earth where there is little or no free oxygen: in the bottoms of swamps—where decaying organic material eats up all the oxygen—and in the guts of some animals, such as people, cows, and termites. And there they have lived in harmony with nature until now.

So what has happened to make their tribe increase? First, the swamps. One particular kind of swamp that has proliferated recently is the rice paddy. As the Third World population drastically increases, it needs more and more food, and thus more and more paddies are cultivated for rice production. In each of these the bacteria find a safe haven, and microscopic bubbles of methane continually burst from the surface of each of them.

Next, the animal guts. The human population is growing, but not enough to seriously increase the numbers of methanogenic bacteria. Cows and termites, however, are another matter. Our appetite for beef in the First World rivals that for rice in the Third, and the cow population grows rapidly under this economic stimulus; with cows come the bacteria, and as each cow flatulates, its intestinal gases are released into the atmosphere, and the methane content increases.

The termites are increasing rapidly in number as a result of another human activity: the burning of rain forests. When forests are opened to human habitation they have to be cleared, and when a peasant stands in a thick growth he doesn't think of buying a bulldozer to clear it; much easier and cheaper just to strike a match and throw it into the forest. This has two unfortunate effects. First, although fire is a thorough leveler of forests, it is also difficult to control and, quite often, too much land is cleared.

The second effect concerns the diet of termites. Beneath each square yard of virgin rain forest there is a termite population of about four thousand critters. (When you think of it, it makes your flesh crawl.) This population, like the population of every species, is kept in check by, among other things, limitations on available food. The food is, of course, wood, and dead wood is preferred to living plants. But what has been recently discovered is that burned wood is easier for the termites to ingest and digest than unburned dead wood. Burning, then, effectively increases the termites' food supply. In the absence of any other ecological change, such as a sudden increase in predators, this leads to an increase in their numbers.

And so the world's methane, like its carbon dioxide, increases. And so the greenhouse effect must certainly increase. Does this mean the world is going to get warmer?

3

Well, yes and no. This is where the answers begin to get vague. Clearly if you wrap another blanket around yourself, you are going to get warmer.

Unless something else happens simultaneously—if your husband turns down the heat, for example, or if he opens the windows and lets in the cold night air. Analogous changes happen on earth that can be organized into two classes. In the first class are events that have occurred in the past, events for which there is definite evidence, although we're not sure exactly what caused them. In the second class are things that people think should, or are afraid might, happen; events that have not yet been seen.

Ice ages, for example, are prime examples of the first class. We have definite evidence that in the past the earth has experienced severe fluctuations in climate, such that the entire top of the world, down to the latitude of New York and Chicago, was covered with ice. This was brought about, almost certainly, by variations in the earth's motion around the sun and around its own axis, although the details are obscure. The time span for such climatic variations is on the order of tens to hundreds of thousands of years, so although more ice ages may occur in the future, they are not an immediate concern. However, there have been smaller variations, documented but largely unexplained, whose temporal characteristics remain a mystery. There was, for example, a period of colder temperatures a thousand years ago.

Toward the end of the ninth century the Vikings settled Iceland,

and a hundred years later, when Eric the Red sailed off to the west, he found an island ringed with greenery, which he named Greenland. A group of Norse families settled there, but were doomed when the weather turned colder.

The following centuries saw a warming trend. In North America the Mill Creek Indian nation established itself in the prairies and valleys east of the Rocky Mountains, but once again the cold came and the weather turned dry, the grassy plains died and could no longer support the buffalo and deer that the Indians hunted, and the stable Indian civilization reverted to the nomadic life it had had before, following the weather and the animals wherever they led.

In Europe there was drought, as much of the water was locked up in northern ice. The Great Fire of London in 1666 was undoubtedly helped by the dryness of the timbers. The Thames would freeze solid from shore to shore in those days, allowing ice-skating parties. But then the climate turned warmer again, into the weather we have today.

The reasons for these warm and cold periods are obscure, but the evidence for them is well founded. So too is the decline in temperature that began in the 1940s. For three decades the globe got cooler year by year, leading some people to fear another slide into a new ice age; you can find articles in magazines like *Time* and *Newsweek* that foretell disaster about to unfold if the temperature continued to dip. But it didn't continue to dip; instead it began to climb, so that the eighties and nineties became the warmest decades on record.

Why did the earth cool off in the middle part of this century, when carbon dioxide levels were still rising? We don't know, but clearly there are more things to climate than are dreamed of in our philosophies. If the dip in temperatures was not caused by carbon dioxide increasing, which certainly it was not, then perhaps—the argument goes—the subsequent increases are also not greenhouse-related.

Perhaps. We do know that carbon dioxide is increasing, and we do know that the earth is getting warmer, but we do not have any direct line of evidence linking cause and effect. But this should not be a reason for complacency: if the recent increases are not due to the greenhouse effect, they will simply make the greenhouse-related increase worse when it kicks in—in the same manner that a hurricane storm surge is at its worst when it comes on top of an unrelated and normal high tide. And since greenhouse gases do trap the infrared light emitted by the earth, their effect *will* have to kick in sooner or later.

In addition to these weather fluctuations that are known to have occurred, there are other fluctuations that ought to occur but for which we lack evidence, at least so far. And to top it all off, it is not *certain* that they ought to occur, because there is so much left unknown.

Take water, for example. Among the host of uncertainties in our knowledge of the future, many are associated with the H_2O molecule in its various manifestations: gaseous vapor, solid ice, and liquid water.

The oceans, to begin with, are composed of liquid water and are probably our largest source of uncertainty. As the atmosphere gets hotter, more and more heat may be absorbed by the surface layer of the ocean. We could calculate fairly well just how much heat the ocean could pull out of the atmosphere in this way, except for the problem of oceanic circulation; that is, if the surface layer *stayed* on the surface, the problem would be straightforward, but it does not stay there. As the waters of the equatorial oceans absorb heat from the hot tropical atmosphere, they get warmer and move toward the poles. Passing into colder regions they give up some of this heat and sink, pulling the remainder of the heat with them. This allows fresh, colder water from the depths to rise near the equator and absorb more heat.

In this simple picture the entire ocean is circulating, and a large portion of the atmospheric heat is sucked to the bottom, where it would remain for a long period of time, probably long enough for our anthropogenic greenhouse effect to be largely washed away. But the actual circulation is much more complex than this, and not as well understood. The warmer the surface layer gets, the less efficiently it will sink to the bottom. In an extreme case, if the entire surface layer gets sufficiently warmer than the deep waters, it would be so buoyant that it would not sink at all. This would effectively stop the circulation of heat down to the oceanic depths, and the amount of heat that can be absorbed by the surface layer alone is not enough to cool off the warming earth. And with all that has been learned about the oceans in recent years, it is still not known what will happen to its circulation patterns as the atmosphere warms. At one extreme, the oceans could act as a heat sink to counterbalance any warming induced by the greenhouse effect; at the other extreme the circulation will be so disturbed that there will be little or no effect.

Solid water, ice, is another problem. The poles of the earth are covered with permanent ice caps, which lock up tremendous amounts of otherwise liquid water and which deflect the sun's rays. That portion of the sun's energy, which otherwise might be absorbed by oceans or solid rock, is bounced away by the ice (particularly by its surface layer of fresh snow) and does not contribute to any warming of the atmosphere; it is reflected away and disappears into space.

If the earth warms, the ice caps may begin to melt and shrink. If that happens, a smaller percentage of the earth's surface will be covered with ice and, obviously, a smaller percentage of the sun's energy will be reflected away. Therefore, a larger percentage of the sun's energy will be absorbed, making the earth even warmer. This is called a positive-feedback effect: a small amount of warming induces a change that results in a larger amount of warming.

Again, it is not known positively that this will happen. A warmer earth will evaporate more water from the oceans, and this water vapor might sail up toward the poles and there condense into snow and ice. This might actually increase the polar caps, and any such increase would reflect away more sunlight than before. That would be a negative-feedback effect: a small change is negated by the change it produces. On the other hand—Harry Truman, it is said, disgusted at always being told two or even three sides of any story by his science advisor, begged God to send him a one-handed scientist—on the other hand, there is another possibility: during the Pliocene period of earth history, 5 million to 10 million years ago, the earth's temperature was just a couple of degrees warmer than now, and a recent study has shown that a large portion of the southern snow cap—the East Antarctic ice sheet—melted. So what will happen tomorrow should the temperature warm up to the Pliocene level? As in the case of oceanic circulation, it is impossible to tell.

There is one more large uncertainty related to water, this time involving its gaseous form, vapor. If more water is evaporated, more clouds will form. A cloudy atmosphere reflects away sunlight just as ice caps do, possibly creating a negative feedback. But clouds are also effective greenhouse agents, trapping infrared radiation coming up from the earth, leading to a positive-feedback effect. Once again, we do not know which way the scales will tip.

The final puzzle that threatens to confound us is methane. And here the latest news is good, though no less puzzling.

In the spring of 1994 the journal *Science* announced that "The rise in . . . methane . . . which had been increasing at a disturbing rate . . . has stopped dead in its track. [And] nobody knows why."

The level is still more than double what it used to be, and it shows no signs of decreasing, but at least it is no longer growing. That's wonderful. But no one understands what happened, and that's not so wonderful. Cow breeding hasn't slowed down, and tropical deforestation hasn't taken much of a beating, and certainly the rate of increase of Third World paddies hasn't ceased. So what's going on?

There are several possibilities. Since *glasnost* the Russians have been busily plugging leaks in their gas lines, and because they're heavily dependent on natural gas fuel there might have been a lot more leakage than anyone suspected. Or the change might be due to the eruption of Mount Pinatubo in 1991 which cooled off the entire earth, giving pause to the warming trends of the past decade. Cooler temperatures might have slowed down natural methane production from bacteria in swamps and paddies.

Or, as *Science* puts it, "Mother Nature may be up to some new tricks" that we don't yet understand.

4

Putting all the uncertainties together we come up with an estimate of the future: the earth will probably warm 1 to 6 degrees Centigrade if the carbon dioxide in the atmosphere doubles, which is likely to happen within the next fifty years. The range in this estimate runs the gamut from slight inconvenience to total calamity. A 1-degree rise in temperature would mean nothing more than a few extra hot nights in Washington and New York and other spots around the globe, while a 6-degree rise would probably melt the ice caps and flood all the coastal cities and plains worldwide, destroy our agricultural system, and spread a host of tropical diseases around the world. Somewhere in the middle range of the estimates, temperatures will be high enough to affect hurricane production and intensity.

And here we run into the remainder of the questions, and the answers get increasingly vague. First of all, while everyone agrees that the world will get warmer in the future, it is not certain how much warmer, nor is it known how the effect will be distributed over the earth. Which regions will get hottest and which will get wettest is anyone's educated guess. So, keeping this in mind for the moment, let's return to the basics.

Hurricanes are powered by the sun's energy, as absorbed in the surface layer of the ocean and subsequently transferred to the atmosphere by evaporation and condensation of water. In order for a hurricane to form, a basic prerequisite is that the ocean surface temperature in the region of formation be greater than about 25 degrees Centigrade (see Figure 4.2, page 94). This is what limits hurricanes to the tropical regions of the world and to the warmer months of the year.

In a greenhouse world the oceans will be warmer. This means that: (1) there will be more days in the year when the tropical waters are warmer than 25 degrees Centigrade, (2) the area in which the maximum temperatures reach this triggering mark will extend farther north (and south, in the southern hemisphere), and (3) there will be more days in which the ocean surface temperature is *much* higher than 25 degrees Centigrade.

Points one and two mean that both the hurricane season and the hurricane regions will be extended: the northern parts of our coast, where hurricanes hardly ever happen today, will be more subject to them tomorrow, and the oceans will reach their triggering temperature earlier in the year and will stay that warm longer into the year. Point three suggests that some hurricanes will form over waters much warmer than the necessary minimum temperature, and so should reach higher wind velocities, causing them to be more vicious. Putting these together suggests that with a longer hurricane season, a greater hurricane-spawning region, and warmer ocean waters, there simply have to be more hurricanes that will cause more damage, even if the probability of a hurricane forming on any one day does not change.

But that probability probably will change. We know that the oceans must reach the triggering temperature in order for a hurricane to have any chance of forming, but we do not know if temperatures warmer than that increase the chance proportionately or if they actually do lead to more destructive storms. Historical analysis shows no greater prevalence of hurricanes in warmer years. For example, the 1980s were the warmest decade on record, but the incidence of hurricanes was much lower than in the cooler 1940s. Still, the statistics are limited, and so is our prognosis. It seems reasonable that if a certain temperature is needed to induce an effect, then the higher you go above that temperature, the easier it ought to be to affect the change and the greater that change ought to be. But this may be overly simplistic, since there are many other conditions necessary for hurricane formation, and some of them may also be affected by the temperature rise, perhaps adversely.

Take the question of localized weather effects, in particular rainfall, and the easterly waves that are the first harbinger of hurricanes (See Chapter 2). We do not know precisely what there is about these waves that leads to the creation of a hurricane, but we do know that that's where they start. We're sort of in the position of

primitive man when he first realized there was a connection between sex and pregnancy: he wasn't clear on just how sex led to
women having babies, but he began to see there was a correlation.

When Hurricane Hugo hit the East Coast and caused $8 billion
worth of damage in 1989, it got William Gray of Colorado State University thinking. For the past several years he had been in the business of predicting the intensity of the coming hurricane season and
had gotten pretty good at it. But in 1989 his predictions hit rock bottom: the year's hurricane-destruction potential more than doubled
his estimate. In particular, no category 4 storm had hit the United
States for quite a while, until Hugo came out of nowhere; he went
back to the records and found that the last one was Camille, in 1969,
but for the twenty years before then a category 4 storm had come
along every four or five years. So he began to think: what had happened to the earth's weather system between 1969 and 1989 that
might account for the lessening in the frequency of intense hurricanes — even though the earth was getting warmer during that period of time?

And one thing stood out: the sub-Saharan drought had begun in
1967. In his predictions Gray had ignored weather over Africa, and
indeed it hadn't seemed to matter. But now he realized that that
might be because for the five years he had been making his predictions, the African weather had always been the same: drought conditions year after year. And then in 1989, the year of Hugo, for the
first time in twenty years, the sub-Saharan rainfall had been normal.

The link is tantalizing, if not exactly clear. We know that the easterly waves flowing off Africa's coast into the Atlantic serve as trig-
gers for hurricanes. It seems reasonable that changing atmospheric
conditions over the African continent will change some characteristics of the easterly waves, so it follows that a drought over Africa
might change the easterly waves to result in an inhibition of hurricane genesis. But that is all that is known at the present time. It is
rather frightening, because droughts there have been cyclical in the
past, and if the cycle is now ending, what is good news for the sub-

Saharan Africans means increased hurricane activity for us in the coming years. And indeed for the years since Hugo, the weather there has been wetter than for the past two decades, the region does seem to be coming out of its drought conditions. The next few years could be a time of nasty hurricane weather.

That is, if the Saharan/hurricane link is a true cause-effect relationship, there are important implications for hurricanes in a greenhouse world. Models of greenhouse warming all predict regional effects, but they don't all agree on what those effects will be. In particular, will the weather over sub-Saharan Africa be wetter in a warmer world, or drier? Will drought conditions come back quickly as the earth warms, or will they be postponed? We simply don't know. Which brings up the last of the questions this chapter started with: what else is likely to happen over the next few decades?

Here the crystal ball blurs over with clouds and fog, with mists and swirls of uncertainties. All the components of the earth's climate system are interrelated in such a complex manner that we are in the position of having to chase after the facts and try to understand what happened whenever something does happen, rather than being able to predict what *will* happen: We're like Monday morning quarterbacks, standing around arguing over why something did this instead of that. Why did storm intensity die down in the years 1969–1989, for example? Is the sub-Saharan drought the entire story? Is it even part of the story, or just a red herring thrown in by God the way Agatha Christie does, to tantalize and confuse? Do Hugo and Andrew signify a return to the higher intensities of the 1947–1969 years, or are they just anomalies? Or will hurricane intensities in the warm decade of the nineties, with the Saharan drought removed, be the worst ever seen?

Although some admit to not knowing the answers, some of us think we do: Dr. Bruce Rosendahl, dean of the Rosenstiel School of Marine and Atmospheric Sciences at the University of Miami, described in *Sea Frontiers* in March of 1993 his "visions for the marine realm" in fifty years: "The [global] warming to date [will have] been sufficient to fuel a string of massively sized hurricanes with 200-mile-per-hour wind speeds. Portions of the Caribbean, Florida, and Gulf Coasts are battered senseless by repeated landfalls. So is the western Pacific."

Other scientists feel differently, and there is no consensus. My own feeling is that it is really very clear that we just do not understand the hurricane sequences of the past fifty, relatively normal years, so how can we possibly predict what will happen in the next fifty, when greenhouse warming becomes a significant but as yet still blurry factor?

5

"Is a puzzlement!"

The words of Oscar Hammerstein with which this chapter began could serve as the national anthem for meteorologists: "In my head are many facts of which I wish I was more certain I was sure." Of course, this is just as true for all scientists, but it strikes more to the heart in meteorology because, for example, as I write these words in the Miami spring of 1993, hurricane season is approaching and I don't know what's going to happen, and with Andrew still reverberating through my house—those damned roof turbines haven't been fixed properly—I'm worried. (I have this out-of-the-future vision: you are reading this book in 1994, shaking your head and musing, The poor guy didn't know what was going to hit him.)

But don't laugh, and ask not for whom the bell tolls. If you live anywhere on the East or Gulf Coasts of the United States, you're in just as much danger, especially as the 1990s pass on and the world gets warmer. And it's not just the coastal areas that are at risk: a good, strong hurricane can move quite a ways inland. The 1938 New England storm moved right on up the Connecticut Valley into Vermont and New Hampshire, maintaining its hurricane intensity all the way. If you live in the Midwestern states you're not much better off, because the same climate effects that can lead to greater hurricanes also have the talent for inducing more and stronger tornadoes, droughts, floods—the entire bestiary of climatic horrors.

Of course, if you live on the West Coast, you have nothing to worry about.

Except earthquakes.

Pleasant dreams.

NOTES

(ALL REFERENCES ARE TO THE BIBLIOGRAPHY.)

PROLOGUE

PAGE

x   *The energy released in:* Nalivkin, p. 52.

x   *In 1954 Hurricane Hazel:* Nalivkin, p. 53.

x   *The Great Hurricane of 1780:* Tannehill (1945), p. 125, quoted in Nalivkin, p. 54.

xi   *The old days were terrible enough:* G. V. Dove, *Zakon Shtormov (Law of Storms),* 1869, quoted in Nalivkin, p. 56.

1:   FIRST ENCOUNTERS

5   *In 1274 Genghis Khan's grandson:* Marshall, p. 220, and Nicolle, p. 65.

7   *"'I have always read,' Columbus wrote":* Letter from Columbus to Ferdinand and Isabella, printed in Lawrence and Young, p. 292.

9   *From what he knew of the dimensions of the land:* Morison (1974), p. 39 and Fisher (1987), p. 22.

10   *In the boats were a dead man and woman:* Morison (1974), p. 16.

11   *That Europeans first saw a hurricane:* Irving, p. 64–69, Vol. II.

12   *They separated into disparate tribes and clans:* Douglas, pp. 39–41.

15   *"Nor were the interests of the unhappy natives":* Irving, p. 95, Vol. II.

15 *When they were enslaved they either fought back or withered away:* E. G. Bourne, *Spain in America,* quoted by M2, p. 175.

17 *The baleful appearance of the heavens:* Irving, p. 311, Vol. II.

17 *Nineteen ships disappeared, with all hands:* Morison (1974), p. 240.

17 *"He was deeply impressed with awe":* Irving, p. 312, Vol. II.

18 *One of the more horrifying stories:* The story was recounted by Aguilar, who was not painted blue and who was allowed to live among the Aztecs. Many years later he was rescued by Cortés, and eventually returned to Spain. Recounted by Douglas, p. 57.

21 *Thus also did they find Bermuda:* Tucker, p. 20.

21 *A contemporary account:* Sylvanus Jordan, *Discovery of the Bermudas, 1610.* Quoted in Tucker, p. 23.

23 *In 1562 Jean Ribaut led an expedition:* Douglas, p. 78.

25 *Among the English sailing baronies:* Douglas, p. 85.

28 *"Over a hundred ships, galleons and merchant ships":* Douglas, p. 106.

28 *"So soone as they had thrown the dead body of the Vice-admiral":* Hakluyt, quoted by Douglas, p. 106.

29 *As the centuries passed:* Douglas, p. 107.

29 *The beginning of true knowledge:* Laughton and Heddon, p. 101.

2: OUT OF NOWHERE

34 *Figure 2-1* from Fisher (1990), p. 25.

46 *In 1821 a hurricane struck New England:* Douglas, p. 224.

46 *But in 1831 an Englishman:* Laughton and Heddon, p. 104.

47 *The most terrible cyclone of modern times:* Tucker, p. 146.

48 *"A knowledge of the law of storms":* Tannehill, p. 26.

48 *"The student navigator is told how to judge":* Tannehill (1944), p. 34.

49 *In the late afternoon of December 17:* Potter, p. 320.

52 *Six months later, off the coast of Okinawa:* Emiliani, p. 283.

55 *Fig. 2-9:* Ahrens, p. 418.

64 *. . . a graduate student at Colorado State University:* Pielke, p. 34.

3: DIRECTION, WHAT DIRECTION?

72 *It was Benjamin Franklin:* Tannehill (1944), p. 5.

74 *It took another ten years: Encyclopedia Brittanica:* "Meteorology"; and Tannehill (1944), p. 5.

74 *"Mr Espy . . . is methodically monomaniacal":* Douglas, p. 225.

78 *Furthermore, the continual accumulation of this converging air:* Pielke, p. 21.

79 *Fig. 3-1* is from Simpson and Riehl, p. 111.

80 *. . . the Great Gale of 1703:* Laughton and Heddon, p. 63.

80 *"They tell us the damage done by the tide":* Defoe, *The Storm,* quoted in Laughton and Heddon, p. 71.

81 *"It's an ill wind that blows no one any good":* Laughton and Heddon, p. 94.

82 *Fig. 3-2* is from Pielke, p. 49.

85 *Instead of blowing steadily along:* Pielke, p. 50.

4: IMPACT

88 *Andrew was to cause more damage:* Total insured losses were 15–16.5 billion dollars. The 1994 Los Angeles earthquake resulted in 2.5 billion dollars in losses. *Facts on File,* 2/14/92.

94 *Fig. 4-2* is from Pielke, p. 42.

95 *Hurricane Damage Scale:* Ahrens, p. 429.

99 *And what rough beast:* From "The Second Coming," by W.B. Yeats.

99 The Bangladesh hurricane of 1970 was described in *Time* and *Newsweek,* November 30, 1970.

100 *SPLASH (Special Program to List Amplitudes of Surge from Hurricanes):* Pielke, p. 71.

100 *Mark Goggin, a Red Cross relief specialist:* From *The Miami Herald:* August 23, 1992.

101 The Galveston, Texas, hurricane of September 7, 1900, is described in Smith, p. 31.

104 The Indianola hurricane is described in Tannehill (1956), p. 65.

107 The Westhampton hurricane is described in *The New York Times,* September 22, 1938, and in Smith, p. 75.

108 *Tex Langford, a twenty-three-year-old cowboy: Time,* October 3, 1938.

110   The 1935 hurricane is described in Nalivkin, p. 53.

110   *A meteorologist at the weather station reported:* Nalivkin, p. 59, quoting Duane, Bulletin of the American Meteorological Society 16, 238–9, 1935.

110   *Among the doctors who hurried down:* Douglas, p. 277.

111   The Pass Christian hurricane is described in *Newsweek*, August 22, 1969, p. 18.

5:   STORM HUNTING

114   *. . . Colonel Joseph P. Duckworth, United States Army Air Force:* Dunn and Banner, p. 156; Tannehill (1956), p. 92ff.

117   *In subsequent flights, aviators found:* Dunn and Banner, p. 163.

120   *"We hit heavy rain and suddenly the airspeed":* Tannehill (1956), p. 176.

121   *"The storm area is approached on a track":* Tannehill (1956), p. 158.

123   *Fig. 5-1* is from Tannehill (1944), p. 23.

125   *The first advances in understanding these storms:* Tannehill (1956), p. 54.

128   *What was happening was this:* Tannehill (1956), p. 70.

132   *". . . one of the great mysteries of the sea":* Tannehill (1956), p. 62.

6:   THE FURIES

139   *All manner of superstition has grown up around the weather:* Byers, p. 4.

142   *If you could harness the energy released:* Ahrens, p. 421.

144   *It was necessary to find an "Achilles' heel" strategy:* Simpson and Riehl, p. 340.

145   The story of Langmuir and Schaefer is taken from Byers, and accounts in *Science* magazine.

153   *Fig. 6-1* is from "Experiments in Hurricane Modification," by R. H. Simpson and J. S. Malkus, in *Scientific American*, December 1964, p. 27.

154   *Fig. 6-2* is from Simpson and Riehl, p. 346.

158   *Years later Simpson was to write:* Simpson and Riehl, p. 342.

165   *An early study of Puerto Rico:* Fassig, quoted in Tannehill (1944), p. 130.

168   *These cooling effects, which have been documented:* Simpson and Riehl, p. 630.

8:   DISASTERS AND WARNINGS

187   *Frank Morkill and his wife, Nancy, in their mid-sixties: The Miami Herald,* August 27, 1992.

188   *Governor Lawton Chiles replied to reporters' questions: USA Today,* August 25, 1992.

191   *"it became a dumping ground for paying off low-level political debts": Newsweek,* September 7, 1992, p. 24.

191   *"FEMA's response to everything is always frustrating": San Francisco Examiner,* October 16, 1992.

194   *Nearly a week later the Red Cross still didn't know: The Miami Herald,* August 29, 1992.

194   *. . . Marlin Fitzwater said it was all Governor Chiles's fault: The Miami Herald,* August 29, 1992.

194   *"Enough is enough," Kate Hale said: The Miami Herald,* September 6, 1992.

194   *U.S. Transportation Secretary Andrew Card: The Miami Herald,* September 6, 1992.

195   *Six months later Henry Cisneros, secretary of Housing and Urban Development: The Miami Herald,* March 4, 1993.

195   *Richard A. Frank, administrator, National Oceanic and Atmospheric Administration:* Baker, p. 5.

196   *"The current state of the art in the evacuation field is not good":* Baker, p. 8.

200   *Fig. 8-2* is from Pielke, p. 198.

202   *Hurricane Celia in 1970:* Simpson and Riehl, p. 299.

202   *"The explosive development of disturbances":* Simpson and Riehl, p. 303.

203   *The models fell into three classes:* Simpson and Riehl, p. 304.

205   *. . . the error in predicting the path of any given storm has been reduced by just about 14 percent:* Kerr, *Science,* February 23, 1990, p. 917.

206   *The Quasi-Lagrangian Model (QLM) of the National Meteorological Center:* Dr. Mukut Mathur, *Hurricane Hugo Conference,* 1993, pp. 301–2, and *Monthly Weather Review,* vol. 119, p. 1419, 1991.

213 *A new system of Doppler radars:* Kerr, *Science,* October 15, 1993.

215 *Satellite measurements can give information on wind speed:* Gray, *Bulletin of the American Meteorological Society,* vol. 72, p. 1871, 1991.

216 *. . . the concept of vertical evacuation:* Baker, p. 16.

218 *. . . when Eloise hit Panama City Beach in 1975:* Baker, p. 16.

218 *After Andrew we found similar crimes in Miami: The Miami Herald,* April 4, 1993.

219 *A second group is suing Walt Disney: The Miami Herald,* April 18, 1993.

9:   FUTURE ENCOUNTERS

235 *The final puzzle that threatens to confound us:* Kerr, *Science,* February 11, 1994.

238 *When Hurricane Hugo hit the East Coast:* Kerr, *Science,* January 21, 1990, p. 162.

239 *. . . "visions for the marine realm":* B. Rosendahl, *Sea Frontiers,* March/April 1993, p. 6.

BIBLIOGRAPHY

Ahrens, C. D. *Meteorology Today.* West, 1985.

Baker, Earl J. (ed.). *Hurricanes and Coastal Storms.* Report No. 33. Florida Sea Grant College: 1980.

Byers, Horace R. "History of Weather Modification," in *Weather and Climate Modification,* W. N. Hess (ed.) New York: Wiley and Sons, 1974.

Douglas, Marjory Stoneman. *Hurricane.* Rinehart & Co., 1958.

Dunn, Gordon E. and Banner, I. Miller. *Atlantic Hurricanes.* Louisiana State University Press, 1964.

Emiliani, Cesare. *A Scientific Companion.* Oxford University Press, 1990.

Fisher, David E. *The Birth of the Earth.* Columbia University Press, 1987.

———. *Fire and Ice.* Harper and Row, 1990.

Hakluyt, R. *Principal Navigations of the English Nation.* London, 1598.

Irving, Washington. *The Life and Voyages of Christopher Columbus.* Putnam, 1856.

Kerr, Richard A. "Hurricane Forecasting Shows Promise." *Science,* February 23, 1990, p. 917.

———. "Hurricane Drought Link Bodes Ill for US Coast." *Science,* January 23, 1990, p. 162.

Laughton, Carr and V. Heddon. *Great Storms.* Philip Allan & Co., 1927.

Lawrence, A.W. and Jean Young. *Narratives of the Discovery of America.* New York: Jonathan Cape and Harrison Smith, 1931.

Marshall, Robert. *Storm from the East.* University of California Press, 1993.

Morison, Samuel Eliot. *The European Discovery of America. The Southern Voyages.* Oxford University Press, 1974.

————. *The Northern Voyages.* 1971.

————. *Admiral of the Ocean Sea.* Little, Brown & Co., 1942.

Nalivkin, D.V. *Hurricanes, Storms and Tornadoes.* Rotterdam: Balkema, 1983.

Nicolle, David. *The Mongol Warriors.* Firebird, 1990.

Pielke, Roger A. *The Hurricane.* Routledge, 1990.

Potter, E.B. *Bull Halsey.* Naval Institute Press, 1985.

Simpson, Robert H. and Herbert Riehl. *The Hurricane and Its Impact.* Louisiana State University Press, 1981.

Smith, Howard E. *Killer Weather.* Dodd, Mead & Company, 1982.

Tannehill, Ivan Ray. *Hurricanes.* Princeton University Press, 1944.

————. *The Hurricane Hunters.* Dodd, Mead & Company, 1956.

Tarbuck, Edward J. and Frederick K. Lutgens. *The Earth.* Merrill, 1984.

Tucker, Terry. *Beware the Hurricane!* Island Press Ltd., 1982.

ABOUT THE AUTHOR

DAVID E. FISHER holds a Ph.D. in nuclear chemistry from the Oak Ridge Institute of Nuclear Studies and the University of Florida. He has been a playwright and actor, and is currently professor of cosmochemistry and director of the Environmental Science Program at the University of Miami. His dozen previous books include both novels and works about science and politics. His last nonfiction book, *Across the Top of the World,* told the history of polar exploration in the context of his journey to the North Pole aboard a Soviet nuclear-powered icebreaker; his last novel, *The Wrong Man,* focused on the resurgence of neo-Nazism. He is married with three children and lives in Miami and East Orleans, Cape Cod.

ABOUT THE TYPE

This book was set in Walbaum, a typeface designed in 1810 by German punch cutter J. E. Walbaum. Walbaum's type is more French than German in appearance. Like Bodoni, it is a classical typeface, yet its openness and slight irregularities give it a human, romantic quality.